NATIONAL ACADEMIES *Sciences Engineering Medicine*

Failure Analysis of the Arecibo Observatory 305-Meter Telescope Collapse

Committee on Analysis of Causes of Failure and Collapse of the 305-Meter Telescope at the Arecibo Observatory

Board on Infrastructure and the Constructed Environment

Division on Engineering and Physical Sciences

Consensus Study Report

NATIONAL ACADEMIES PRESS 500 Fifth Street, NW Washington, DC 20001

This activity was supported by Grant CMMI- 2135084 from the National Science Foundation to the National Academy of Sciences. Any opinions, findings, conclusions, or recommendations expressed in this publication do not necessarily reflect the views of any organization or agency that provided support for the project.

International Standard Book Number-13: 978-0-309-70222-5
International Standard Book Number-10: 0-309-70222-4
Digital Object Identifier: https://doi.org/10.17226/26982

Cover: Photo courtesy of the Arecibo Observatory, a facility of the National Science Foundation.

This publication is available from the National Academies Press, 500 Fifth Street, NW, Keck 360, Washington, DC 20001; (800) 624-6242 or (202) 334-3313; http://www.nap.edu.

Copyright 2024 by the National Academy of Sciences. National Academies of Sciences, Engineering, and Medicine and National Academies Press and the graphical logos for each are all trademarks of the National Academy of Sciences. All rights reserved.

Printed in the United States of America.

Suggested citation: National Academies of Sciences, Engineering, and Medicine. 2024. *Failure Analysis of the Arecibo Observatory 305-Meter Telescope Collapse*. Washington, DC: The National Academies Press. https://doi.org/10.17226/26982.

The **National Academy of Sciences** was established in 1863 by an Act of Congress, signed by President Lincoln, as a private, nongovernmental institution to advise the nation on issues related to science and technology. Members are elected by their peers for outstanding contributions to research. Dr. Marcia McNutt is president.

The **National Academy of Engineering** was established in 1964 under the charter of the National Academy of Sciences to bring the practices of engineering to advising the nation. Members are elected by their peers for extraordinary contributions to engineering. Dr. John L. Anderson is president.

The **National Academy of Medicine** (formerly the Institute of Medicine) was established in 1970 under the charter of the National Academy of Sciences to advise the nation on medical and health issues. Members are elected by their peers for distinguished contributions to medicine and health. Dr. Victor J. Dzau is president.

The three Academies work together as the **National Academies of Sciences, Engineering, and Medicine** to provide independent, objective analysis and advice to the nation and conduct other activities to solve complex problems and inform public policy decisions. The National Academies also encourage education and research, recognize outstanding contributions to knowledge, and increase public understanding in matters of science, engineering, and medicine.

Learn more about the National Academies of Sciences, Engineering, and Medicine at **www.nationalacademies.org**.

Consensus Study Reports published by the National Academies of Sciences, Engineering, and Medicine document the evidence-based consensus on the study's statement of task by an authoring committee of experts. Reports typically include findings, conclusions, and recommendations based on information gathered by the committee and the committee's deliberations. Each report has been subjected to a rigorous and independent peer-review process and it represents the position of the National Academies on the statement of task.

Proceedings published by the National Academies of Sciences, Engineering, and Medicine chronicle the presentations and discussions at a workshop, symposium, or other event convened by the National Academies. The statements and opinions contained in proceedings are those of the participants and are not endorsed by other participants, the planning committee, or the National Academies.

Rapid Expert Consultations published by the National Academies of Sciences, Engineering, and Medicine are authored by subject-matter experts on narrowly focused topics that can be supported by a body of evidence. The discussions contained in rapid expert consultations are considered those of the authors and do not contain policy recommendations. Rapid expert consultations are reviewed by the institution before release.

For information about other products and activities of the National Academies, please visit www.nationalacademies.org/about/whatwedo.

**COMMITTEE ON ANALYSIS OF CAUSES OF FAILURE AND COLLAPSE OF
THE 305-METER TELESCOPE AT THE ARECIBO OBSERVATORY**

ROGER L. McCARTHY (NAE), McCarthy Engineering, *Chair*
RAMÓN L. CARRASQUILLO,[1] Carrasquillo Associates
DIANNE CHONG (NAE), Boeing Engineering, Operations & Technology (retired)
ROBERT B. GILBERT (NAE), The University of Texas at Austin
W. ALLEN MARR, JR. (NAE), Geocomp, Inc.
JOHN R. SCULLY, University of Virginia
SAWTEEN SEE, See Robertson Structural Engineers
HABIB TABATABAI, University of Wisconsin–Milwaukee

Study Staff

CAMERON OSKVIG, Board Director, Study Director
JAYDA WADE, Research Associate (until July 31, 2023)
JOSEPH PALMER, SR., Program Assistant
RADAKA LIGHTFOOT, Financial Business Partner (until March 20, 2023)
DONAVAN THOMAS, Financial Business Partner (from March 20, 2023)

---

[1] Deceased on February 2, 2024.

BOARD ON INFRASTRUCTURE AND THE CONSTRUCTED ENVIRONMENT

JESUS M. DE LA GARZA, Clemson University, *Chair*
BURCU AKINCI, Carnegie Mellon University
STEPHEN AYERS, Ayers Group
BURCIN BECERIK-GERBER, University of Southern California
LEAH BROOKS, The George Washington University
MIKHAIL V. CHESTER, Arizona State University
JAMES (JACK) DEMPSEY, Asset Management Partnership, LLC
LEONARDO DUENAS-OSORIO, Rice University
DEVIN K. HARRIS, University of Virginia
DAVID J. HAUN, Haun Consulting, Inc.
CHRISTOPHER J. MOSSEY, Fermi National Accelerator Laboratory
ANDREW PERSILY, National Institute of Standards and Technology
ROBERT B. RAINES, Atkins Nuclear Secured
JAMES RISPOLI, North Carolina State University
DOROTHY ROBYN, Boston University
SHOSHANNA D. SAXE, University of Toronto

Staff

CAMERON OSKVIG, Board Director
JAMES MYSKA, Senior Program Officer
BRITTANY SEGUNDO, Program Officer
JOSEPH PALMER, SR., Senior Program Assistant
DONAVAN THOMAS, Finance Business Partner

Dedication

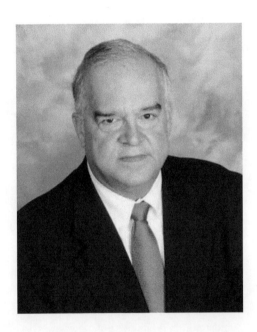

This report is dedicated to committee member Dr. Ramón L. Carrasquillo, who unexpectedly passed away before this report's release. His pragmatic and insightful contributions strengthened the report. In addition to his extensive engineering and materials science expertise, his deep connection to Puerto Rico helped the committee develop a nuanced understanding of the community and culture surrounding the Arecibo Observatory. He is remembered by the committee as a thoughtful, warm, and generous colleague.

NOTE: Image courtesy of Carrasquillo Associates.

Reviewers

This Consensus Study Report was reviewed in draft form by individuals chosen for their diverse perspectives and technical expertise. The purpose of this independent review is to provide candid and critical comments that will assist the National Academies of Sciences, Engineering, and Medicine in making each published report as sound as possible and to ensure that it meets the institutional standards for quality, objectivity, evidence, and responsiveness to the study charge. The review comments and draft manuscript remain confidential to protect the integrity of the deliberative process.

We thank the following individuals for their review of this report:

DONALD CAMPBELL, Cornell University
GREGORY G. DEIERLEIN (NAE), Stanford University
LENNARD FISK (NAS), University of Michigan
DAVID GOODYEAR (NAE), Independent Consultant
MARTHA HAYNES (NAS), Cornell University
LT. COL. (RET.) CLARENCE (BART) KEMPER, Kemper Engineering Services, LLC
MATTHYS LEVY (NAE), Thornton Tomasetti
MOHAMMAD MODARRES, University of Maryland
JANINE PARDEE, Independent Consultant
RANDALL POSTON (NAE), Pivot Engineers

Although the reviewers listed above provided many constructive comments and suggestions, they were not asked to endorse the conclusions or recommendations of this report nor did they see the final draft before its release. The review of this report was overseen by **WILLIAM F. BAKER,** Skidmore Owings and Merrill, LLP, and **STEVE BATTEL (NAE),** Battel Engineering. They were responsible for making certain that an independent examination of this report was carried out in accordance with the standards of the National Academies and that all review comments were carefully considered. Responsibility for the final content rests entirely with the authoring committee and the National Academies.

Contents

PREFACE xiii

SUMMARY 1

1 INTRODUCTION 7
History of the Arecibo Telescope, 7
Arecibo Telescope Cable System, 12
Statement of Task, 16

2 THE COLLAPSE: WHAT HAPPENED 18
Arecibo Telescope Failure Sequence, 18
Hurricane Maria Hits the Arecibo Telescope, 18
Post-Maria Arecibo Telescope Inspections, 22
The Hurricane Maria Aftermath, 26
Bureaucratic Delays in Funding Arecibo Telescope Hurricane Repairs, 28
Sequence of Cable Failure Events, 30

3 ANALYSIS 35
Cable Socket Zinc Creep Failure, 35
Cable End Sockets, 48
Wire Breaks, 49
Earthquake, 55
Wind Speed Consideration in the Arecibo Telescope's Design, 56
Governing Cable Design Standards, 56
Arecibo Telescope Cable Load, 58
Risk Considerations, 61
Structural Robustness, 63
Monitoring, 65

4	**ARECIBO TELESCOPE'S MANAGEMENT AND OVERSIGHT**	**70**
5	**OTHER LESSONS LEARNED** State of Knowledge, 74 Continued Research, 75	**74**

BIBLIOGRAPHY   76

APPENDIXES

A	Committee Member Biographical Information	81
B	Information-Gathering Activities	84
C	Arecibo Telescope Cable Failure Mechanisms Considered by the Committee	86
D	Arecibo Telescope Design Issues Considered by the Committee	94
E	Acronyms and Abbreviations	96

Preface

It has been my privilege to chair this committee of distinguished subject-matter experts in its investigation and final probable cause determination of one of the most publicized and baffling failures of the modern era. It became clear shortly after the Arecibo Telescope's collapse that the zinc used to anchor the steel supporting cable wires into their sockets had allowed the failed cables to slip out of their sockets, known as spelter sockets. The sockets slowly lost their grip on a critical number of the cable wires via slow zinc "creep," a process where the zinc deformed slowly at a load below half the socket's nominal strength. Although the committee agrees with the conclusions from other forensic reports regarding zinc creep at the connection being the failure mechanism, the baffling question was, "Why was there excessive zinc creep at such loading?" Such a failure had never been reported previously in over a century of widespread zinc spelter socket successful use.

Fortunately, the committee had the benefit of the detailed analysis and well-documented reports from NASA; Wiss, Janney, Elstner Associates, Inc.; and Thornton Tomasetti, Inc., without which it could not have completed its task. Building on their work, the committee presents a clear and plausible explanation of why the telescope's sockets failed when no such sockets have ever been reported to have failed before. Unfortunately, there was not enough data available to prove our explanation. It is simply the most plausible hypothesis based on the data we do have.

Without the depth and breadth of expertise on the committee, its task would remain uncompleted. I think I speak for everyone on the committee when I say that none of us could have done this alone. I want to thank my colleagues for their unwavering dedication to the task. Their professionalism and competence made my job an enjoyable one.

Roger L. McCarthy, *Chair*
Committee on Analysis of Causes of Failure and Collapse
of the 305-Meter Telescope at the Arecibo Observatory

Summary

The Committee on Analysis of Causes of Failure and Collapse of the 305-Meter Telescope at the Arecibo Observatory of the National Academies of Sciences, Engineering, and Medicine was asked to review the failure and collapse of the 305-Meter Telescope at the Arecibo Observatory in Puerto Rico and explain the contributing factors and probable cause(s) of the failure, as well as recommendations for measures to prevent similar damage to other facilities in the future.

After analyzing the data and the extensive and detailed forensic investigations commissioned by the University of Central Florida and the National Science Foundation (NSF),[1] the committee consensus is that the root cause of the Arecibo Telescope's collapse was unprecedented and accelerated long-term zinc creep induced failure of the telescope's cable spelter sockets. The finding of creep aligns with the other forensic investigations. "Each failure involved both the rupture of some of the cable's wires and a deformation of the socket's zinc, and is therefore the failure of a cable-socket assembly."[2] While the cable system was upgraded in 1997 with cable safety factors greater than two,[3] the telescope collapsed after the failure of several cable-socket assemblies that were not loaded at the time of the first socket failure above half their nominal design strength.[4] Failure despite a factor of safety of 2 was possible because the accelerated time-dependent materials failure process governing eventual zinc pull out (i.e., power law creep [PLC]) occurred at stresses below 50 percent of the cable strength.

Throughout the investigation, the committee looked for an explanation for the non-uniform accelerated zinc creep that led to the three socket pullout failures that led to its collapse from the Arecibo Telescope's population

---

[1] The Wiss, Janney, Elstner Associates (WJE) report on the initial auxiliary socket and cable (WJE, 2021, *Auxiliary Main Cable Socket Failure Investigation*, WJE No. 2020.5191, June 21); NASA Engineering and Safety Center (NESC) and Kennedy Space Center report (G.J. Harrigan, A. Valinia, N. Trepal, P. Babuska, and V. Goyal, 2021, *Arecibo Observatory Auxiliary M4N Socket Termination Failure Investigation*, NASA/TM–20210017934, NESC-RP-20-01585, NASA Engineering and Safety Center, Langley Research Center, June 15, https://ntrs.nasa.gov/api/citations/20210017934/downloads/20210017934%20FINAL.pdf), and forensic investigations performed by Thornton Tomassetti, Inc. (TT)—the "TT Interim Report" (TT, 2021, *Arecibo Telescope Collapse: Forensic Investigation Interim Report*, NN20209, prepared by J. Abruzzo and L. Cao, November 2) and the "TT Final Report" (TT, 2022, *Arecibo Telescope Collapse: Forensic Investigation*, NN20209, prepared by J. Abruzzo, L. Cao, and P.E. Pierre Ghisbain, July 25, https://www.thorntontomasetti.com/sites/default/files/2022-08/TT-Arecibo-Forensic-Investigation-Report.pdf).

[2] TT Final Report, p. 1.

[3] TT Final Report, Appendix C, Table 4, p. 9.

[4] TT Final Report, p. 188, Figure 25.

of 78[5] zinc-filled sockets. In over a century of successful use prior to the Arecibo Telescope's collapse, all the forensic investigations agreed that such a spelter socket failure had never been reported. The committee considered the following unanswered questions about the Arecibo Telescope's collapse:

1. Why did the Arecibo Telescope sockets/cables fail despite their widely used safety factor above 2? (the ratio of cable strength to applied load)
2. Why did this unprecedented spelter socket accelerated zinc creep failure mode appear in the Arecibo Telescope and nowhere else in more than a century of successful spelter socket use?
3. In the Arecibo Telescope environment, why did a relatively young auxiliary socket, M4N-T, fail first, long before any main cable socket with more than twice as much age with three-plus decades of previous service (1963–1997) at an even lower safety factor?
4. Why did four of the six platform auxiliary cable sockets have no further pullout beyond the ⅜ inch they gained during installation (as measured in the Lehigh University socket test)[6] and thus exhibited zero measured creep even after 23 years of service?
5. Why did all six auxiliary cable tower sockets exhibit cable pullout beyond their initial ⅜ inch, indicating some creep after installation, and why did all the auxiliary cables exhibit more socket pullout on the tower end than the platform end of the same cable?
6. Why did the rate of wire breakage in the main cables decrease significantly after 1974?
7. Why did the auxiliary cables have no recorded wire breaks even after 23 years of service?

The only hypothesis the committee could develop that provides a plausible but unprovable answer to all these questions and the observed socket failure pattern is that the socket zinc creep was unexpectedly accelerated in the Arecibo Telescope's uniquely powerful electromagnetic radiation environment. The Arecibo Telescope cables were suspended across the beam of "the most powerful radio transmitter on Earth."[7] The other investigations failed to note several failure patterns and provided no plausible explanation for most of them. To answer these questions with empirical evidence instead of only the inferences that can be made from the existing data, a more comprehensive and widespread forensic analysis of "good" and "bad" socket workmanship and the low-current, long-term effect on zinc creep is required. This report makes the recommendation to provide this evidence using the remaining socket and cable sections recovered from the site.

Recommendation: While still available, the National Science Foundation should offer the remaining socket and cable sections to the research community for continued fundamental research on large-diameter wire connections, the long-term creep behavior of zinc spelter connections, and materials science.

In preparing this report, the committee first assessed "what happened" and then provided analysis as to the most probable explanation it could develop with the available evidence as to "why it happened." The committee gathered important information from the Arecibo Telescope's administrators, engineers, and technicians, as well as NSF, as well as expertise from bridge and structural engineers. The committee also reviewed the in-depth forensic reports performed after the failure and reviewed the structural analysis, engineering plans, inspection reports, and photographs of the Arecibo Telescope, as well as the various repair proposals. Finally, a literature review was performed of zinc spelter sockets, cable connection pullout, zinc creep, static and dynamic loading of zinc, and electric current effects on zinc creep. In addition, due to the technical complexity of this report, before the report was finalized, NSF was provided an opportunity to review portions of the text and make suggestions relating to any perceived technical inaccuracies or factual errors.

The committee concluded that a 39-month failure sequence (Figure 2-1), starting with Hurricane Maria (hereafter "Maria") on September 20, 2017, then a Category 4 storm, led to the telescope's collapse. Structural analysis

---

[5] TT Final Report, p. 46.

[6] TT Final Report, Appendix N.

[7] A.P.V. Siemion, et al., 2011, "Developments in the Radio Search for Extraterrestrial Intelligence," *2011 XXXth URSI General Assembly and Scientific Symposium*, https://doi.org/10.1109/URSIGASS.2011.6051263.

performed as part of the forensic investigations established that the wind loading from the hurricane should not have damaged the telescope's cable structure or caused any additional socket cable pullout. The structural analysis demonstrated that Maria's wind loading did not increase any cable tension below a nominal safety factor of 1.8.[8] However, the failure sequences occurred at cable loads well below the traditional factor of safety through some mechanism(s) not previously reported. The sequence likely began with Maria because inspections that were performed on the Arecibo Telescope sockets in 2003 and 2011 by Ammann & Whitney (A&W), a structural engineering firm, reported that "the end socket ½" cast zinc leading edge separation was observed at all cables but has not measurably increased since reported in the 2003 survey."[9] The best available information would indicate cable pullouts remained on the order of ½ inch until after Maria. As a result of sparse inspection documentation, there is uncertainty around how much damage was evident immediately post-Maria. Inspections made in late 2018 and early 2019 observed cable slippage greater than 1.5 inches on auxiliary sockets at the ground end of backstay B12W and 1.125 inches at the tower end of M4N. "The cable slips were evidence of structural distress in the sockets and should have raised a concern that cables may fail."[10] The question is why cable slip did not spark greater concern. It might be that the gravity of the situation was not recognized for the following two reasons: (1) The traditional safety factor based on cable strength divided by cable loads was still comfortably above 2 depending on the particular cable, and (2) in structures using spelter connections, the weak link has never been the connector itself.

The Arecibo Telescope gave fair warning post-Maria that it was in structural distress through increasing cable socket pullout. Upon reflection, the unusually large and progressive cable pullouts of key structural cables that could be seen during visual inspection several months and years before the M4N failure should have raised the highest alarm level, requiring urgent action. The lack of documented concern from the contracted engineers about the inconsequentiality of cable pullouts or the safety factors between Maria in 2017 and the failure is alarming. Given the observed excessive cable pullout, continued use of the factor of safety calculations based on the original design ignored the impact of any degradation mechanism. Safety factors based on cable strength are not pertinent to failure modes involving creep, stress corrosion cracking, hydrogen embrittlement, fatigue, or the integrity of the socket connections.

These mechanisms routinely occur at applied stresses below half of the yield strength in a material undergoing one of these processes. Unprecedented structural distress evidence in any of NSF's massive structures should not be casually dismissed. The risks posed by the structural distress should have been alarming to inspectors, taking into consideration the safety and lives of the facility's personnel. Despite the safety factors, these risks merited an urgent response.

Nothing in the ASCE 19[11] or AASHTO M 277[12] standards indicates that any movement (pullout) of the cables from their sockets after initial installation (approximately ⅜ inch) was acceptable or justifiable. The structural analysis demonstrated that even Maria wind loading should not have added significant incremental stress to affect the safety factor nor occurred over a time period great enough to affect PLC or cause additional cable pullout. The reliance by the consultants (before and after the first cable failure in 2020) on a perceived allowable pullout of one-sixth of the cable diameter, which should only be seen at loading at 80 percent of ultimate cable strength, does not align with the AASHTO M 277 standard guidance. The committee, therefore, disagrees with the suggestion made in the Thornton Tomasetti, Inc. (TT) 2022 report, *Arecibo Telescope Collapse: Forensic Investigation*,[13] to use the D/6 limit as a threshold for slip monitoring.

Almost 3 years after Maria, on August 10, 2020, the tower end of auxiliary cable 4, labeled M4N-T, at less than half its design load, pulled out of its zinc-filled spelter socket and failed. The loose cable struck the Gregorian dome and crashed onto the dish below. At the time of failure, it had only been in service 23 years after the 1997

---

[8] TT Final Report, Appendix J, Figure 34, p. 35.
[9] Ammann & Whitney, 2011, "Arecibo Radio Telescope Structural Condition Survey," Cornell University Archives, Arecibo Ionospheric Observatory Records #53-7-3581, Division of Rare and Manuscript Collections, Cornell University Library Box 37, Folder 8, March, p. 2.
[10] TT Interim Report, p. 25.
[11] Refers to the American Society of Civil Engineers standard "Structural Applications of Steel Cables for Buildings," ASCE 19.
[12] Refers to the American Association of State Highway and Transportation Officials standard "Standard Specification for Wire Rope and Sockets for Movable Bridges," AASHTO M 277.
[13] TT Final Report, p. 49.

upgrade. The main cables had seen roughly 57 years of service by August 2020 with no socket failures at that point in time, with at least 30 of those years (1963–1993) at a ~10 percent higher static dead-load safety factor (1.98 as opposed to 2.18) than the auxiliary cables saw before the M4N-T failure. The socket that failed first, M4N-T, was not the most heavily loaded, nor was the brooming (the physical spread of the cable wires before being cast in the molten zinc, Figure 1-6[14]) of this socket's wires the worst found among the Arecibo Telescope's socket connections.

After the M4N-T failure, the main cables to Tower 4, M4 (×4), were loaded to 646 kips,[15] compared to 516 kips in the M12 (×4) mains and 497 kips in the M8 (×4) mains, as illustrated in Figure 2-7. This re-distribution of loads reduced the TT-computed (but incorrect) safety factor for the M4 (×4) mains to 1.6, making the M4 (×4) mains significantly more heavily loaded compared to their original strength than any other remaining cables on the telescope. Thus, while the cables should not have failed even at this load, it is not surprising that, given the evident socket degradation, the second and third failures were in the M4 (×4) mains.

On November 6, 2020, one of the M4 (×4) main cables on Tower 4, M4-4, failed again at the zinc-filled spelter socket (Figure 2-7). Interim repairs were set to begin on November 9. The material was already being staged when the second cable pulled out. After this second cable failure, the loads on the three remaining M4 (×3) cables increased to 800 kips, reducing the erroneously determined safety factor to 1.3[16] for the remaining M4 cables On November 19, NSF announced that safe repairs would not be possible and that Arecibo Telescope would be closed in a controlled decommissioning. On December 1, M4-2, one of the three remaining zinc-filled sockets holding the M4 (×3) cables on Tower 4, failed, increasing the load in the two remaining M4 (×2) cables to 1,062 kips above their nominal strength. The 913-ton platform collapsed, swung across the dish, and smashed through the reflector.

Absent Maria, the committee believes the telescope would still be standing today but might have eventually collapsed if its unique, accelerated zinc creep had not been addressed before it was decommissioned. The long-term zinc creep failure of the Arecibo Telescope sockets and the subsequent cable pullout has never been documented elsewhere despite a century of zinc-filled cable spelter sockets use.[17] The type, size, length, and fittings of the cables used in the Arecibo Telescope (whether the original cables constructed in the 1960s or the auxiliary cables installed in the 1990s) were catalog-selected items, not at all unusual, with decades of proven performance. Extensive structural modeling of the Arecibo Telescope, independently confirmed by laser cable sag surveys, validated that under all static and cyclic loading conditions, the cable loads barely exceed half the nominal cable strength. While PLC can occur at stresses well below yield, such a creep failure has never been reported in spelter socket zinc.

The Arecibo Telescope's cable design adhered to the standards of practice during the original design and the subsequent addition of the Gregorian dome in 1997. Accounting for static load, cyclic dynamic loading from thermal effects, wind, and earthquakes, and operating in a corrosive tropical environment, the telescope's engineering design and material specifications were reasonable for the original construction and subsequent upgrades. Construction procedures and workmanship were adequate to ensure long-term structural integrity. Extensive investigation and testing revealed no defects in the Arecibo Telescope's design, materials, or workmanship that contributed materially to its collapse. The committee did not find sufficient evidence or analysis pointing to an unrecognized design fault, defects in the socket construction, other environmental effects such as hydrogen embrittlement, or some unobserved dynamic load condition based upon connection design and platform geometry. However, these factors cannot be fully discounted. As stated previously, a more comprehensive and widespread forensic analysis of both "good" and "bad" socket workmanship and mechanisms to accelerate zinc creep would be required.

A potential mechanism for spelter socket zinc creep acceleration not considered in other analyses was the effect of low-current electroplasticity (LEP). The cables whose sockets failed were suspended in a unique and powerful radio telescope environment, capable of inducing current in the cables at some level. Electric current flowing through zinc has been found to increase its creep rate but under laboratory conditions significantly different

---

[14] Brooming refers to the physical spread of the cable wires before being cast in the molten zinc. Refer to Figure 1-6.

[15] A kilopound (kip) is a non-metric unit of force equal to 1,000 pounds-force. 1 kip is equivalent to 4448.2216 Newton.

[16] TT Final Report, Appendix G, Figure 22, p. 17.

[17] "Wire Cables of Various Types and Materials Tested by U.S. Bureau of Standards," 1915, *Engineering Record* 72(19):567–568.

than the spelter socket service in the Arecibo Telescope.[18] While there is not enough data or empirical evidence to prove LEP as a causal mechanism for the acceleration of the socket zinc creep, no other mechanism has been found likely. The circumstantial evidence and the cable pullout patterns offer support for the role of LEP. LEP provides a physically plausible but unproven mechanism to answer the outstanding questions described above about why this spelter socket failure mode was seen in the Arecibo Telescope and nowhere else in history, the non-uniformity in the rate and pattern of the cable pullouts, the failure of a young auxiliary cable socket first, and the timing of the Arecibo Telescope's cable wire breaks.

All the reported experimental zinc electroplasticity (EP) data were developed at current densities orders of magnitude higher than those possibly present in the Arecibo Telescope but measured in laboratory experimental periods that were orders of magnitude shorter than the telescope's socket zinc service. There are no reported experimental data concerning low-current, long-term EP, which the committee has lumped together under the term "LEP," affecting zinc's creep mechanisms over decades. There was also no reported measuring capability or data from the Arecibo Telescope concerning induced currents or electromagnetic effects in the cables. Measurements of induced current in the Arecibo Telescope's cables at peak broadcast power or the quality of the various tower and platform grounding paths were not recorded and thus could not empirically validate this explanation. The timing and patterns of the Arecibo Telescope's socket failures make the LEP hypothesis the only one that the committee could find that could potentially explain the failure patterns observed. This mode of accelerated PLC would have also been operative at stresses below the strength of the cable.

By necessity, NSF's cutting-edge research will occasionally require unique custom-designed facilities that place conventional structural designs and materials into new and unprecedented operating regimes where prior experience in conventional environments is not a reliable guide. Unprecedented failure modes in NSF's other large facilities can never be fully anticipated. But to the extent reasonable, their onset and progression can and should be detected by careful condition monitoring of performance, which must be increased and not decreased with age. TT observed, "The available information [about the Arecibo Telescope's condition] is generally less comprehensive and detailed after the second upgrade in 1997. The scope of the inspections performed by A&W was considerably reduced, and the quarterly maintenance reports were replaced with simple maintenance logs."[19] For aged structures such as the Arecibo Telescope, additional facility maintenance and monitoring (and their associated costs) may be warranted. The committee does not know how much of this monitoring and inspection reduction was caused directly or indirectly by NSF's reduction in Arecibo funding over its final decade of service. The committee concluded that the safety consequences of a structural failure of the Arecibo Telescope were not considered in decision-making during its design and operation or in decisions about extending its life.

This report makes recommendations for critical facilities to have formal facility operation and inspection manuals and routine independent monitoring and assessment.

Recommendation: The facility owner/operator should ensure that an operations and maintenance manual for the structure is commissioned and is available during the operation of the structure. The manual should:
- **Identify performance standards of the facility to help detect unexpected, potentially dangerous performance and deteriorating performance with time;**
- **Provide a monitoring and inspection plan that considers potential critical failure modes (and necessary inspection expertise to address them) and include physical variables to monitor, locations to monitor, and the recommended frequency of monitoring. The plan should recognize that some time-dependent failure modes can operate at low loads in contradiction with the safety factor. It should also provide limit values for warning levels and action levels for each performance variable to be monitored. (Warning level is the point where performance becomes concerning, and further evaluation of the safety of the structure should be made. The limit level**

---

[18] A. Lahiri, P. Shanthraj, and F. Roters, 2019, "Understanding the Mechanisms of Electroplasticity from a Crystal Plasticity Perspective," *Modelling and Simulation in Materials Science and Engineering* 27(8), https://doi.org/ARTN 085006 10.1088/1361-651X/ab43fc.
[19] TT Interim Report, p. 6.

endurance limit is the point where the performance is becoming threatening to life, and people should be removed from harm's way.); and
- Indicate the expected service life of the facility and its key components.

Recommendation: The facility owner/operator should:
- **Implement the monitoring plan and keep it operational for the life of the structure. For structures with long life expectancies, this plan may require updating to account for mechanisms and degradation that are a function of age; and**
- **Engage a qualified professional to evaluate the monitoring data at least annually, assess the safety of the structure, and provide recommendations for changes to the structure and changes to the monitoring plan as needed.**

This higher level of monitoring and analysis is much more likely to happen if NSF makes explicit funding provisions for detailed condition maintenance and monitoring that do not effectively operate as unbudgeted contractor costs or penalties. These funds need to be accompanied by mechanisms for enforcing regular inspections, monitoring, maintenance, and repair. The decision to decommission a major structural facility should be an orderly and safe process. Reduced funding for maintenance and monitoring runs the risk that nature will "decommission" some facilities, such as the Arecibo Telescope, by processes that are inevitably disastrous, and may include loss of life.

Recommendation: The National Science Foundation and organizations that use similar site management contracts to manage their portfolios should consider funding for the inspection, monitoring, maintenance, and repair of aging facilities and infrastructure as important as they are critical to the structure's performance and longevity.

With respect to oversight of contractor operated facilities, the committee made the following recommendation.

Recommendation: Facility owners should enforce compliance with contract requirements through independent auditing of inspection, monitoring, maintenance, and repair records.

1

Introduction

The Arecibo Telescope was a 1,000-foot (305-meter) fixed spherical radio/radar telescope and the primary instrument of the National Astronomy and Ionosphere Center, also known as the Arecibo Observatory (AO). The Arecibo Telescope was completed in 1963 and was the world's largest single-aperture radio telescope until 2016. It was used mainly for research in space sciences, atmospheric sciences, and radio and radar astronomy. With the telescope, the AO made important scientific contributions, such as the creation of the first radar maps of the surface of Venus, the first discovery of a binary pulsar, the first discovery of an exoplanet, and numerous observations of Earth's ionosphere. Although initial funding for the Arecibo Telescope's design and construction came from military sources, the National Science Foundation (NSF) became the government agency monitoring the Arecibo contract beginning in 1967. The AO was managed by Cornell University from 1963 to 2011.[1]

HISTORY OF THE ARECIBO TELESCOPE

Construction of the Arecibo Telescope occurred from 1960 to 1963. A photograph of the telescope at its completion in August 1963 is shown in Figure 1-1.[2] The telescope had a cable-suspended 1,220 kilopound (kip)[3] (~610 ton [short]) feed platform that supported a rotatable truss. The platform was suspended almost 500 feet above the 1,000-foot diameter spherical reflector ("dish") from three towers, with each tower connected to the platform by four ~3 inch diameter, ~575 feet long main cables, resulting in a total of 12 main cables.[4] These three groups of four main cables were attached to each of the three support towers, as shown. The feed platform's loading on the towers was counterbalanced by five ~3.25 inch diameter backstay cables on each tower that were, in turn, attached

---

[1] National Science Foundation (NSF), "Arecibo: Facts and Figures," https://www.nsf.gov/news/special_reports/arecibo/Arecibo_Fact_Sheet_11_20.pdf, accessed June 1, 2023.

[2] M. Zastrow, 2021, "The Rise and Fall of Arecibo Observatory—An Oral History," *Astronomy.com*, https://www.astronomy.com/science/the-rise-and-fall-of-arecibo-observatory-an-oral-history.

[3] Thornton Tomassetti, Inc. (TT), 2022, *Arecibo Telescope Collapse: Forensic Investigation*, NN20209, prepared by J. Abruzzo, L. Cao, and P.E. Pierre Ghisbain, July 25, https://www.thorntontomasetti.com/sites/default/files/2022-08/TT-Arecibo-Forensic-Investigation-Report.pdf (hereafter "TT Final Report"), p. 1.

[4] G.J. Harrigan, A. Valinia, N. Trepal, P. Babuska, and V. Goyal, 2021, *Arecibo Observatory Auxiliary M4N Socket Termination Failure Investigation*, NASA/TM–20210017934, NESC-RP-20-01585, NASA Engineering and Safety Center, Langley Research Center, June 15, https://ntrs.nasa.gov/api/citations/20210017934/downloads/20210017934%20FINAL.pdf (hereafter "NESC Report"), p. 18.

FIGURE 1-1 The Arecibo Telescope shown as commissioned in 1963.
SOURCE: Cornell University, courtesy of AIP Emilio Segrè Visual Archives.

to a backstay anchorage.[5] When the Arecibo Telescope was completed in 1963, its radar system was reported to be a "430 MHz system, pulsed transmitter, 2.5 MW peak, 150 kW maximum average power, which enabled the lunar radar mapping at 70 cm wavelength."[6]

The Arecibo Telescope received two major upgrades in its lifetime that made significant changes to its structure and equipment. The first upgrade of the telescope was completed in 1974. It replaced the mesh surface of the telescope's dish reflector with perforated aluminum panels, allowing for higher-frequency observations. The world's largest radar transmitter was also upgraded with a new S-band radar consisting of a 2,380 MHz continuous wave (CW) transmitter, with a 2.5 MW peak and 450 kW average power.[7] The upgrade allowed for S-band transmitting capabilities to improve the radar imaging resolution of the surface of Venus and other objects. The telescope's platform after this first upgrade is shown in Figure 1-2.[8]

---

[5] NESC Report.

[6] T.W. Thompson, B.A. Campbell, and D.B.J. Bussey, 2016, "50 Years of Arecibo Lunar Radar Mapping," *URSI Radio Science Bulletin* 2016(357):23–35, https://doi.org/10.23919/URSIRSB.2016.7909801, p. 27.

[7] *Microwave Journal*, Volume 25, May 1982, pp. 111, 112, 114 (4 ff.).

[8] TT Final Report, Appendix A, Figure 4, p. 4.

FIGURE 1-2 The Arecibo Telescope's original feed platform in 1982.
SOURCE: Thornton Tomasetti, 2022, *Arecibo Telescope Collapse: Forensic Investigation*, NN20209, prepared by J. Abruzzo, L. Cao, and P.E. Pierre Ghisbain, July 25, https://www.thorntontomasetti.com/sites/default/files/2022-08/TT-Arecibo-Forensic-Investigation-Report.pdf; modified from photo by Manfred Niermann, Wikipedia, CC BY-SA 4.0; courtesy of Thornton Tomasetti.

Between the first upgrade and the second, the Arecibo Telescope's structure did not change significantly. Only one noteworthy event occurred between upgrades: the replacement of a backstay cable in 1981. A sixth broken wire was found at the ground end of this cable, pushing the observatory to replace the cable to prevent further wire breaks and potential cable failure.[9] Wire breaks are discussed in more detail in Chapter 3.

The Arecibo Telescope received a second upgrade, completed in 1997, that significantly changed its suspended structure and cable system. This upgrade involved the addition of the Gregorian dome to the rotatable truss (azimuth arm) below the suspended platform and a second line feed for the ionospheric radar.[10] The Gregorian dome provided secondary and tertiary reflectors that corrected the dish's spherical aberration and provided a multi-beam receiver.[11] A more powerful 2,380 MHz CW, 1.0 MW maximum power, radar was installed.[12] Since the Gregorian dome added significant weight to the structure—it was five times heavier than the removed carriage house—a counterweight was added to the azimuth arm opposite the Gregorian dome. Finally, 12 auxiliary cables were added to the original cable system. In total, the upgraded suspended platform weighed approximately 1.8 million pounds (~900 tons [short]). In addition, the inclined tie-downs were replaced with vertical tie-downs and linear actuators (jacks) to minimize the elevation fluctuation of the suspended structure during daily temperature cycles. The addition of these linear actuators increased the cable tensions slightly. A log of the tiedown forces recorded from 2004 indicated that the total tiedown force increased by an average of 60 kips at night under the combined effect of temperature and jack pulldown, which increased the tensions in the entire cable system. The original structure was only equipped with passive tiedown cables, and therefore, the impact of day-night cycles on cable tensions was less significant. "The combined effect of the additional tiedown tension and temperature drop is a 3 percent increase in the main and backstay cable tensions."[13]

---

[9] TT Final Report, Appendix C, p. 4.
[10] Wiss, Janney, Elstner Associates (WJE), 2021, Auxiliary Main Cable Socket Failure Investigation, WJE No. 2020.5191, June 21 (hereafter "WJE Report"), p. 2.
[11] P.A. Taylor and E. Rivera-Valentín, 2021, "Fall of an Icon: The Past, Present, and Future of Arecibo Observatory," *Lunar and Planetary Institute Information Bulletin*, Issue 165.
[12] T.W. Thompson, et al., 2016, "50 Years of Arecibo Lunar Radar Mapping," p. 27.
[13] TT Final Report, p. 28.

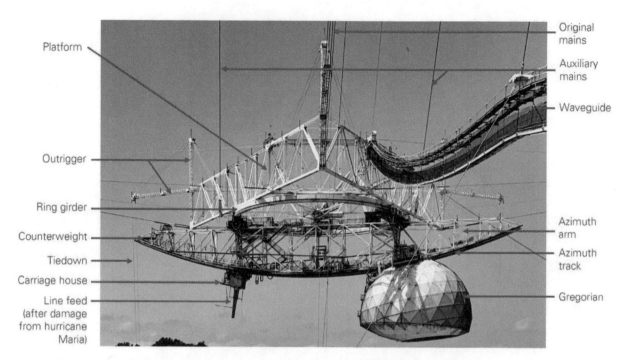

FIGURE 1-3 The Arecibo Telescope's suspended platform in 2019, post–Hurricane Maria.
SOURCE: Thornton Tomasetti, 2022, *Arecibo Telescope Collapse: Forensic Investigation*, NN20209, prepared by J. Abruzzo, L. Cao, and P.E. Pierre Ghisbain, July 25, https://www.thorntontomasetti.com/sites/default/files/2022-08/TT-Arecibo-Forensic-Investigation-Report.pdf; modified from photo by Mario Roberto Durán Ortiz, Wikipedia, CC BY-SA 4.0; courtesy of Thornton Tomasetti.

After this upgrade, the Arecibo Telescope remained substantially unchanged until its collapse in 2020. Its suspended platform and overall configuration after the 1997 upgrade are shown in Figure 1-3[14] and Figure 1-4,[15] respectively.

In 2006, NSF released its Astronomical Science Senior Review Committee Report[16] analyzing NSF's astronomy facilities. The report contained a recommendation to close the AO by 2011 unless other sources could be found to fund its operation.[17] In 2011, NSF awarded management of the facility to SRI International with an NSF-reduced annual budget of $8 million and an additional $3.6 million provided by NASA. NSF proposed to reduce its contribution to $6.08 million (a reduction of 24 percent) by fiscal year (FY) 2019 in NSF's FY2019 Budget Request to Congress.[18] On April 1, 2018, the University of Central Florida (UCF) consortium officially took over the operation of the AO facility, with the commitment that "by October 1, 2022, NSF's contribution will shrink to $2 million per year, with the UCF consortium making up the difference."[19] "Since Fiscal Year 2018, NSF has contributed around $7.5 million-per-year to Arecibo operations and management."[20]

Months before UCF took over the operation of the AO, in September 2017, Hurricane Maria, then a Category 4 hurricane,[21] struck Arecibo. While the AO's weather station produced the only local available

---

[14] TT Final Report, Appendix A, p. 5.

[15] TT Final Report, Appendix A, Figure 1-5, p. 2.

[16] NSF, 2006, "From the Ground Up: Balancing the NSF Astronomy Program," Report of the National Science Foundation Division of Astronomical Sciences Senior Review Committee, October 22, https://www.nsf.gov/mps/ast/seniorreview/sr_report_mpsac_updated_12-1-06.pdf (hereafter "NSF Senior Review").

[17] NSF Senior Review, p. 6.

[18] NSF, 2018, "FY 2019 NSF Budget Request to Congress," February 28, https://new.nsf.gov/about/budget/fy2019, p. Facilities-7.

[19] D. Clery and A. Cho, 2018, "Iconic Arecibo Radio Telescope Saved by University Consortium," *Science*, February 22, https://www.science.org/content/article/iconic-arecibo-radio-telescope-saved-university-consortium.

[20] NSF, "Arecibo: Facts and Figures," p. 1.

[21] H. Weitering, 2017, "Hurricane Maria Damages Parts of Puerto Rico's Arecibo Observatory," *Space.com*, September 22, https://www.space.com/38242-arecibo-observatory-hurricane-maria-damage.html.

FIGURE 1-4 Arecibo Telescope in its post-1997 upgrade configuration.
SOURCE: Thornton Tomasetti, 2022, *Arecibo Telescope Collapse: Forensic Investigation*, NN20209, prepared by J. Abruzzo, L. Cao, and P.E. Pierre Ghisbain, July 25, https://www.thorntontomasetti.com/sites/default/files/2022-08/TT-Arecibo-Forensic-Investigation-Report.pdf; modified from photo by the National Science Foundation; courtesy of Thornton Tomasetti.

wind speed data, differing analyses have estimated differing peak wind speeds of 105 mph,[22] 108 mph,[23] 110 mph,[24] 110 mph,[25] 110 mph,[26] and 118 mph.[27] Most major structures of the Arecibo Telescope stood intact, but some of the damage included the loss of one of the line feeds on the antenna for one of the radar systems, as well as punctures in the telescope's dish.[28] Two million dollars were awarded in the summer of 2018, 9 months after Hurricane Maria, to UCF to be focused on repairs judged to be the most time-critical.[29] The bulk of the funding ($12.3 million) for repairs was awarded in the summer of 2019;[30] these funds did not cover Tower 4 cable replacement, which had not been identified as a critical need, but did include "Tower 8 spliced main cable replacement."[31]

---

[22] WJE Report, p. 6.
[23] TT Final Report, p. 12.
[24] WJE Report, p. 43.
[25] TT Final Report, Appendix J, p. 1.
[26] NESC Report, p. 100.
[27] TT Final Report, Appendix J, p. 25.
[28] S. Farukhi, 2017, "Latest USRA Update on Arecibo Observatory—September 22, 2017," Newsroom, Universities Space Research Association, September 22, https://newsroom.usra.edu/latest--usra--update-on-arecibo-observatory.
[29] A. VanderLey, 2022, "Arecibo Observatory: Failure Event Sequence," National Science Foundation presentation to the committee, January 25 (hereafter "NSF presentation"), slide 13.
[30] NSF presentation, slide 14.
[31] NSF presentation.

On August 10, 2020, one of the auxiliary cables on Tower 4 pulled out of its socket. It struck the Gregorian dome and crashed onto the dish below.[32] On November 6, another cable, one of the four main cables on Tower 4, failed.[33] NSF announced its decision to decommission the Arecibo Telescope on November 19,[34] and they stated that it would be closed in a controlled decommissioning. Finally, on December 1, the remaining cables on Tower 4 failed, and the suspended platform collapsed, smashing through the reflector.[35] The collapse of the telescope is discussed in further detail in Chapter 2.

ARECIBO TELESCOPE CABLE SYSTEM

Since the cable system was at the center of the Arecibo Telescope's collapse, a brief description of the cables and overall system is included here. The telescope's platform was initially suspended by 12 main cables—4 for each of the 3 towers—and 5 backstay cables attached to each tower to a backstay anchorage behind the tower. In the 1997 upgrade, 12 auxiliary cables were added to the original 12 main cables—for each tower, 2 additional 713-foot-long auxiliary main cables were added to deal with the 40 percent subsequent increase in platform weight (now totaling 913 tons)[36] resulting from the dome addition,[37] as well as 2 additional auxiliary backstay cables from each tower to react to the additional loading from the new auxiliary platform cables.[38] The auxiliary platform cables were isolated and did not run in parallel with the main cables. The main cables were attached to the triangular corners of the platform, and the newly installed auxiliary cables were attached along the respective reinforced transverse steel trusses. Arecibo Telescope's cable nomenclature and geometry are illustrated and labeled in both the plan view shown in Figure 1-5[39] and the side view shown in Figure 1-6.[40]

The physical properties of the five different cable types that suspended the Arecibo Telescope after the 1997 upgrade are illustrated in Figure 1-7.[41] As Figure 1-7 illustrates, an auxiliary main cable weighs 22 lb/ft, giving the 713-foot-long cable[42] a total weight of ~15.7 kips.

Each of these cables had zinc-filled spelter sockets at both ends.[43] A spelter socket is a steel block with a cone-shaped cavity where the cable is inserted. The wires of the cable are spread out before the cavity is filled with molten zinc.[44] A diagram of the cable "socketing" and pre-stretching is shown in Figure 1-8.

"The cable-socket assembly is not expected to fail before the cable. Samples of the auxiliary main and backstay cables were tested in 1993 before the second upgrade of the telescope. For both samples, the first wire ruptures occurred several feet away from the sockets and under a load higher than the cable's Minimum Breaking Strength."[45] These zinc-filled spelter sockets are the key elements in the Arecibo Telescope's failure and are discussed extensively in Chapter 3.

In engineering, the "safety factor" of a structure is the ratio of its strength to an intended load. More specifically, in cable structural design, the safety factor is "defined as the cable's minimum breaking strength divided by its actual tension."[46] The Arecibo Telescope's original cable system had an average safety factor of approximately 2.0 under the self-weight of the telescope, and 1.67 considering a 140-mph wind speed in addition to the dead

---

[32] M. Williams, 2020, "An Update on the Damage to the Arecibo Observatory," *Universe Today*, September 14, https://www.universetoday.com/147782/an-update-on-the-damage-to-the-arecibo-observatory.

[33] TT Final Report, p. 13.

[34] NSF presentation, slide 38.

[35] TT Final Report, p. i.

[36] J. Abruzzo, L. Cao, and P. Ghisbain, 2022, "Arecibo Observatory: Stabilization Efforts and Forensic Investigation," Thornton Tomassetti, Inc., presentation to the committee, February 17, slide 12.

[37] TT Final Report, p. 2.

[38] TT Final Report, Appendix C, p. 8.

[39] TT Final Report, Appendix A, Figure 7, p. 7.

[40] TT Final Report, Appendix A, Figure 8, p. 7.

[41] TT Final Report, Appendix A, Figure 1-9, p. 9.

[42] WJE Report, p. 15.

[43] TT Final Report, p. 5.

[44] TT Final Report.

[45] TT Final Report, p. 5.

[46] TT Final Report, p. 6.

INTRODUCTION

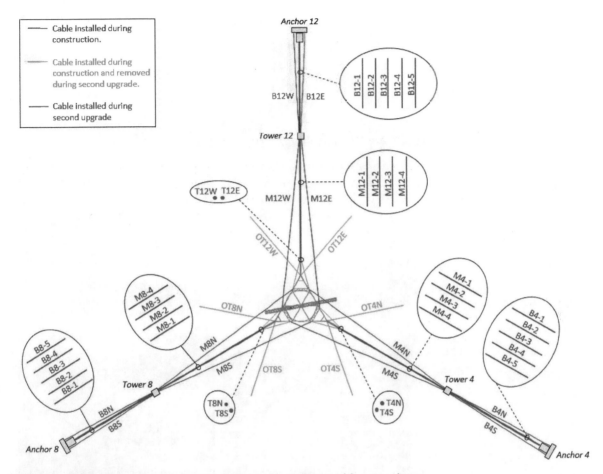

FIGURE 1-5 Plan view of the Arecibo Telescope's cable geometry with nomenclature.
SOURCE: Thornton Tomasetti, 2022, *Arecibo Telescope Collapse: Forensic Investigation*, NN20209, prepared by J. Abruzzo, L. Cao, and P.E. Pierre Ghisbain, July 25, https://www.thorntontomasetti.com/sites/default/files/2022-08/TT-Arecibo-Forensic-Investigation-Report.pdf; courtesy of Thornton Tomasetti.

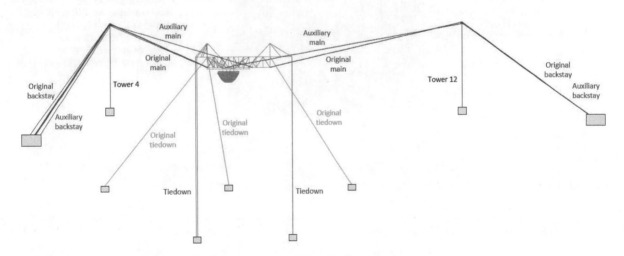

FIGURE 1-6 Side view of the Arecibo Telescope's post-1997 cable geometry.
SOURCE: Thornton Tomasetti, 2022, *Arecibo Telescope Collapse: Forensic Investigation*, NN20209, prepared by J. Abruzzo, L. Cao, and P.E. Pierre Ghisbain, July 25, https://www.thorntontomasetti.com/sites/default/files/2022-08/TT-Arecibo-Forensic-Investigation-Report.pdf; courtesy of Thornton Tomasetti.

Original main
Cable diameter = 3 in
Wire diameter = 13/64 in
Number of wires = 168
Linear weight[1] = 19 lbf/ft
Min. breaking strength[2] = 1,044 kip

Auxiliary main
Cable diameter = 3 1/4 in
Wire diameter = 1/4 in
Number of wires = 126
Linear weight[1] = 22 lbf/ft
Min. breaking strength[2] = 1,314 kip

Original backstay
Cable diameter = 3 1/4 in
Wire diameter = 13/64 in
Number of wires = 216
Linear weight[1] = 22 lbf/ft
Min. breaking strength[2] = 1,212 kip

Replacement B12-3
Cable diameter = 3 1/4 in
Wire diameter = 1/4 in
Number of wires = 126
Linear weight[1] = 22 lbf/ft
Min. breaking strength[3] = 1,314 kip

Auxiliary backstay
Cable diameter = 3 5/8 in
142 wires of 1/4 in diameter
23 wires of 3/16 in diameter
Linear weight[1] = 28 lbf/ft
Min. breaking strength[2] = 1,614 kip

[1] Typical linear weight per ASTM A586.
[2] Per 1992 structural drawings for second upgrade.
[3] Assumed identical to auxiliary mains.

FIGURE 1-7 Illustration of the Arecibo Telescope's five cable types after the 1997 upgrade.
SOURCE: Thornton Tomasetti, 2022, *Arecibo Telescope Collapse: Forensic Investigation*, NN20209, prepared by J. Abruzzo, L. Cao, and P.E. Pierre Ghisbain, July 25, https://www.thorntontomasetti.com/sites/default/files/2022-08/TT-Arecibo-Forensic-Investigation-Report.pdf; courtesy of Thornton Tomasetti.

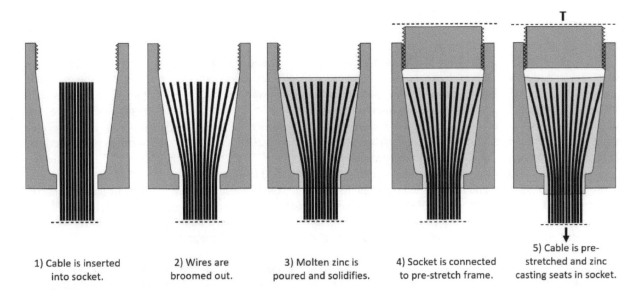

FIGURE 1-8 Arecibo Telescope cable socketing and pre-stretching.
SOURCE: Thornton Tomasetti, 2022, *Arecibo Telescope Collapse: Forensic Investigation*, NN20209, prepared by J. Abruzzo, L. Cao, and P.E. Pierre Ghisbain, July 25, https://www.thorntontomasetti.com/sites/default/files/2022-08/TT-Arecibo-Forensic-Investigation-Report.pdf; courtesy of Thornton Tomasetti.

load. Following the telescope's second upgrade, the average safety factor of the cable system was 2.25 under dead load and 2.15 during the design windstorm, which specified a wind speed of 110 mph.[47]

In addition to maintenance operations done on the telescope's cable system by AO staff, the Engineer of Record Amman & Whitney (A&W) conducted structural inspections between 1972 and 2011. However, the scope of these inspections was reduced after the telescope's 1997 upgrade.[48] A&W noted in both of its 2003 and 2011 inspections of the Arecibo Telescope auxiliary main cables that "cable slip" had occurred in the auxiliary main and backstay cables: "the cast zinc of the spelter sockets had separated from the leading edge of the sockets by up to ½."[49] As described earlier, ⅜ inch of cable pullout was most likely due to fabrication and proof loading. A&W merged with and operated under the name Louis Berger, U.S., in 2016, which was later acquired by WSP Global, Inc., in 2018.

The 2006 NSF Astronomical Science Senior Review Committee Report[50] explained, "The National Astronomy and Ionosphere Center (NAIC) is operated by Cornell University and runs the Arecibo Observatory. Its current budget is $12M comprising $10M from NSF/AST and $2M from NSF/Division of Atmospheric Sciences (ATM)."[51] It contained a recommendation to close the Arecibo Observatory by 2011 unless other sources could be found to fund its operation.[52] In 2011, NSF awarded management of the facility to SRI International with an NSF-reduced annual budget of $8 million and an additional $3.6 million provided by NASA. NSF proposed to reduce its contribution to $6.08 million (a reduction of 24 percent) by FY 2019 in NSF's FY2019 Budget Request to Congress.[53] On April 1, 2018, the UCF consortium officially took over the operation of the AO facility, with

---

[47] TT Final Report, p. 7.
[48] TT Final Report, p. 8.
[49] Ammann & Whitney, 2011, *Arecibo Radio Telescope Structural Condition Survey*, Cornell University Archives, Arecibo Ionospheric Observatory Records #53-7-3581, Division of Rare and Manuscript Collections, Cornell University Library Box 37, Folder 8, March, Section 3.2.4, p. 3.
[50] NSF, 2006, "From the Ground Up."
[51] NSF Senior Review.
[52] NSF Senior Review, p. 6.
[53] NSF, 2018, "FY 2019 NSF Budget Request to Congress," February 28, https://new.nsf.gov/about/budget/fy2019, p. Facilities-7.

the commitment that "by October 1, 2022, NSF's contribution will shrink to $2 million per year, with the UCF consortium making up the difference."[54]

In September 2017, Hurricane Maria, then a Category 4 hurricane,[55] struck Arecibo. Afterward, "The major structures including the 300-m telescope were intact, though suffered some damage when the atmospheric radar line feed broke off and falling debris from it punctured the dish in several places. A separate 12-m dish used as a phase reference for Very Long Baseline Interferometry was lost."[56] Also, "A 96-foot (29-m) antenna that was suspended above the telescope's 1,000-foot (305-m) dish was lost as a result of the hurricane."[57] The Arecibo Telescope was inspected and $2 million were awarded in the summer of 2018 to UCF to be focused on repairs judged to be the most time-critical.[58] Repairs to the transmitter were ongoing in 2020 when the COVID-19 pandemic hit.

On August 10, 2020, one of the auxiliary cables, M4N-T, attached to the top of Tower 4, pulled out of its socket. It struck the Gregorian dome and crashed onto the dish below.[59] On November 6, one of the four main cables on Tower 4 failed.[60] On November 19, NSF announced its decision to decommission the Arecibo Telescope[61] and that it would be closed in a controlled decommissioning. On December 1, the remaining cables on Tower 4 failed, and the 913-ton platform[62] collapsed, swung across the dish, and smashed through the reflector.[63]

The two cable failures that occurred before the collapse and the third cable failure that triggered the collapse all happened near or within zinc-filled spelter sockets at cable ends. No cable failed away from a socket before the collapse. Each failure involved the rupture of some of the cable's wires and a deformation of the socket's zinc and is, therefore, the failure of cable-socket assembly.[64]

While the cable system was designed with safety factors greater than 2, the telescope collapsed after the failure of several cable-socket assemblies.[65]

STATEMENT OF TASK

This report fulfills the statement of task reprinted in Box 1-1. The committee relied on laboratory testing, analyses, and the excellent graphics and illustrations in three forensic reports. These reports were conducted by Wiss, Janney, Elstner Associates, Inc., *Auxiliary Main Cable Socket Failure Investigation—Final Report* (2021),[66] NASA Engineering and Safety Center *Arecibo Observatory Auxiliary M4N Socket Termination Failure Investigation* (2021),[67] and Thornton Tomasetti, Inc., *Arecibo Telescope Collapse: Forensic Investigation* (2022).[68] In addition to these reports, the committee gathered information from engineering research literature as well as presentations from NSF staff, AO staff, the forensic investigation teams, and structural engineering experts. The committee reviewed inspection reports and photographs and Arecibo Telescope engineering design and repair documents. Lastly, National Academies policy and procedure allows for technical or descriptive portions of text to be reviewed for accuracy by the sponsor, relevant organization, or a content expert. Due to the technical complexity of this report, before the report was finalized, NSF was provided an opportunity to review the text and suggest corrections of any perceived technical inaccuracies or factual errors.

---

[54] D. Clery and A. Cho, 2018, "Iconic Arecibo Radio Telescope Saved by University Consortium," *Science*, February 22, https://www.science.org/content/article/iconic-arecibo-radio-telescope-saved-university-consortium.
[55] Weitering, 2017, "Hurricane Maria Damages Parts of Puerto Rico's Arecibo Observatory."
[56] Farukhi, 2017, "Latest USRA Update on Arecibo Observatory—September 22, 2017."
[57] Weitering, 2017, "Hurricane Maria Damages Parts of Puerto Rico's Arecibo Observatory."
[58] NSF presentation, slide 13.
[59] Williams, 2020, "An Update on the Damage to the Arecibo Observatory."
[60] TT Final Report, p. 13.
[61] NSF presentation, slide 38.
[62] TT Final Report, p. i.
[63] TT Final Report.
[64] TT Final Report, p. 1.
[65] TT Final Report, p. 15.
[66] WJE Report.
[67] NESC Report.
[68] TT Final Report.

> **BOX 1-1**
> **Statement of Task**
>
> At the request of the Director of the National Science Foundation, the National Academies of Sciences, Engineering, and Medicine will convene an ad hoc study committee to conduct a review of the failure and collapse of the 305-Meter Telescope at the Arecibo Observatory in Puerto Rico. The committee will issue a report explaining the contributing factors and probable cause(s) of the failure and recommendations for measures to prevent similar damage to other facilities in the future. The committee will assess the environmental, physical, and design considerations as well as any administrative or management practices that may have been contributing factors to the failure and include the following tasks:
>
> 1. Examine the performance of the structures related to:
>   a. Engineering design and material specification for original and subsequent upgrades;
>   b. Documented construction procedures and contractor performance;
>   c. Environmental conditions, loading events (e.g., wind, seismic, multi-hazard), corrosion;
>   d. Maintenance, repair, and recapitalization activities for the telescope;
> 2. Assess oversight and management policies and practices that may have been contributing factors to the failure, including:
>   a. Contractor selection and procurement during construction and repair;
>   b. Maintenance planning and oversight;
>   c. Routine inspection and structural review;
> 3. Identify lessons learned for NSF in general for oversight and response actions for other large facilities physical condition, integrity, and function, including end-of-life considerations; and
> 4. Identify and recommend actions or general best practices for consideration to limit or prevent other large facility engineering failure or damage at other large NSF facilities that would significantly impact ongoing science.

2

The Collapse: What Happened

ARECIBO TELESCOPE FAILURE SEQUENCE

The more than 3-year sequence of selected events that led to the ultimate failure of the Arecibo Telescope is illustrated chronologically in Figure 2-1. As can be seen from this timeline, the final cable failures themselves that caused the Arecibo Telescope's collapse comprise only 10 percent of the more than 3-year Arecibo Telescope failure sequence timeline.

HURRICANE MARIA HITS THE ARECIBO TELESCOPE

The Arecibo Telescope's failure sequence timeline begins when Hurricane Maria hit the Arecibo Telescope on September 20, 2017, almost 39 months before the Arecibo Telescope's collapse. Major natural disasters, including hurricanes, earthquakes, and floods, represent a significant challenge to the integrity of aging structures, especially those designed and built decades before the disaster and which have reached or exceeded their original design service life. Among others, these challenges include the predictability of their magnitudes or forces, orientation of the events with respect to the structure, location, topography, incidence, occurrences, and redundancy of the structure. The Arecibo Telescope survived many natural disasters, including multiple earthquakes, one of which was 6.4 magnitude, and hurricanes. The most significant was Hurricane Maria. The committee could not discern exactly what goals, guidelines, and/or instructions were followed by the contracted and onsite engineers and inspectors that inspected the Arecibo Telescope after each natural disaster, and it is not clear that any "red lines" or critical thresholds were established. As discussed below, no reference or owner's manual was ever prepared for the facility by the original designers or the engineers involved in the upgrades to guide a series of contract operators on the risks or inspections of structurally critical components.

Maria subjected the Arecibo Telescope to winds of between 105 and 118 mph, with the source of this uncertainty in wind speed discussed below. After every natural disaster, the Arecibo Telescope was at least visually inspected by the staff and sometimes by outside engineers, as it was after Maria. Although the Arecibo Telescope exhibited evidence of the effects of these natural disasters, every inspection concluded that no significant damage had jeopardized the Arecibo Telescope's structural integrity.

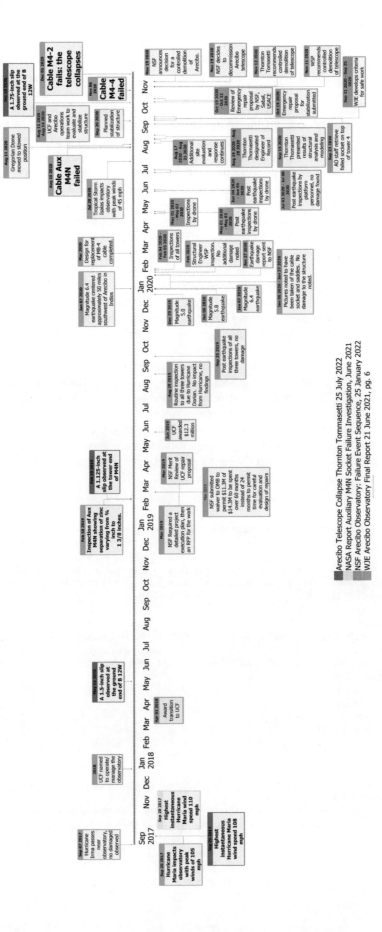

FIGURE 2-1 The Arecibo Telescope's failure sequence timeline.

Based on a review of available records, the winds of Hurricane Maria subjected the Arecibo Telescope's cables to the highest structural stress they had ever endured since it opened in 1963.[1] While the Arecibo Telescope weather station produced the only available local wind speed data, varying analyses have estimated differing peak wind speeds of 105 mph,[2] 108 mph,[3] 110 mph,[4,5,6] and 118 mph.[7] The variance in reported wind speed is dependent on the time interval of the measurement (3-sec, 15-sec, 1-min, 2-min, 10-min, or "fastest mile") or is a reporting of the mean value of the measurements. Additional discussion of reported wind speed is in Appendix D. The Wiss, Janney, Elstner Associates (WJE) and Thornton Tomasetti, Inc. (TT) reports (hereafter "WJE Report" and "TT Final Report," respectively) present different estimates of the "peak" or "highest instantaneous" wind speed seen by the Arecibo Telescope. These differing estimates of the peak wind speed at the Arecibo Telescope during Maria result from a concurrent Arecibo Telescope weather instrument failure. The previously cited peak wind speed estimates are based on the same Arecibo Telescope Doppler data recorded at 15-second intervals.[8] Arecibo Telescope's telescope platform was equipped with a weather station that measured wind speed with two independent instruments.[9] Unfortunately, one of these instruments, the anemometer, whose measurements of wind speed were logged every second,[10] "failed approximately one hour before the peak of the storm,"[11] so the 15-second interval Doppler data has been used to estimate what the actual peak wind speed in between the 15-second measuring intervals would have been. Since the Arecibo Telescope's structural "wind load is proportional to the square of the wind speed and to the structure's area modified by a drag coefficient,"[12] going from a 105 mph to 118 mph peak wind speed increases the calculated peak drag forces from Hurricane Maria by more than 25 percent.

The maximum wind speed the Arecibo Telescope was designed to withstand, and with what factor of safety, is unclear from the records and has changed with upgrades. The National Aeronautics and Space Administration (NASA) observed, "It is worth reflecting on the original design that considered the "survival" condition to be 100 mph winds."[13] After the Gregorian dome upgrades, the "survival loading" condition included wind effects from 100 mph winds.[14] However, TT reported, "The structural drawings for the original structure indicate a design wind speed of 140 mph."[15] TT reports that after the Gregorian upgrade, "A&W confirmed the two speeds of 110 mph and 123 mph for the global and local design of the upgrade respectively and, to our knowledge, the final design was based on those wind speeds."[16] A summary of the available documentation is illustrated in Figure 2-2.[17]

Because of the high wind speeds noted above, TT devoted an entire appendix to wind loading on the Arecibo Telescope from Maria.[18]

---

[1] Wiss, Janney, Elstner Associates (WJE), 2021, *Auxiliary Main Cable Socket Failure Investigation*, WJE No. 2020.5191, June 21 (hereafter "WJE Report"), p. 16.

[2] WJE Report, p. 6.

[3] Thornton Tomasetti, Inc. (TT), 2022, *Arecibo Telescope Collapse: Forensic Investigation*, NN20209, prepared by J. Abruzzo, L. Cao, and P.E. Pierre Ghisbain, July 25, https://www.thorntontomasetti.com/sites/default/files/2022-08/TT-Arecibo-Forensic-Investigation-Report.pdf (hereafter "TT Final Report"), p. 12.

[4] WJE Report, p. 43.

[5] TT Final Report, Appendix J, p. 1.

[6] G.J. Harrigan, A. Valinia, N. Trepal, P. Babuska, and V. Goyal, 2021, *Arecibo Observatory Auxiliary M4N Socket Termination Failure Investigation*, NASA/TM–20210017934, NESC-RP-20-01585, NASA Engineering and Safety Center, Langley Research Center, June, https://ntrs.nasa.gov/api/citations/20210017934/downloads/20210017934%20FINAL.pdf (hereafter "NESC Report"), p. 100.

[7] TT Final Report, Appendix J, p. 25.

[8] TT Final Report, Appendix J, p. 26.

[9] TT Final Report, Appendix J, p. 6.

[10] TT Final Report, Appendix J, p. 8.

[11] TT Final Report, Appendix J, p. 25.

[12] TT Final Report, Appendix J, p. 3.

[13] NESC Report, p. 99.

[14] WJE Report, p. 34.

[15] TT Final Report, Appendix J, p. 1.

[16] TT Final Report, Appendix J, p. 2.

[17] J. Abruzzo, L. Cao, and P. Ghisbain, 2022, "Arecibo Observatory: Stabilization Efforts and Forensic Investigation," Thornton Tomassetti, Inc., presentation to the committee, February 17 (hereafter "TT presentation"), slide 34.

[18] TT Final Report, Appendix J.

THE COLLAPSE: WHAT HAPPENED

The highest instantaneous wind speed ever recorded at Arecibo Observatory is 110 mph and occurred as the eye of Hurricane Maria approached the site on September 20, 2017. While the hurricane caused widespread destruction in Puerto Rico, the only significant damage reported on the telescope was a failure of the line feed. However, because of the relatively short time between Hurricane Maria and the telescope's first cable failure (August 10, 2020), it is natural to investigate the storm as a potential factor contributing to the collapse. Hurricane Maria is therefore used as a reference load throughout the analysis of the wind load effects.[19]

TT estimated Arecibo Telescope cable stresses from Hurricane Maria winds assuming a basic wind speed of 110 mph.[20] It was estimated in the analysis that, based on an assumption of a 110-mph peak windspeed, the tension in the M4N cable increased 10 percent from the Maria wind loading. TT concluded, "Two recent extreme events—Hurricane Maria in 2017 and the January 2020 earthquake sequence—were specifically analyzed and estimated to have temporarily increased the cable tensions by up to 15 percent."[21] But even with the additional hurricane wind load, the analysis found the M4N cable to be the second (631 kips) most lightly loaded of all the auxiliary cables during Maria[22] and was therefore calculated to have the highest safety factor (a tie with 2.1 "in hurricane conditions") of any auxiliary cable on the Arecibo Telescope.[23]

Design Wind Speed

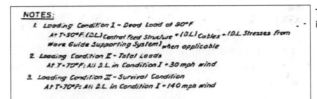

The **original** structural drawings (12/1/1960) indicate a design wind speed of **140 mph**.

The structural drawings for the **first upgrade** (12/30/1972) indicate a design wind speed of **110 mph**.

The structural drawings for the **second upgrade** (1992) indicate a design wind speed of 100 mph.

However, correspondence between Ammann&Whitney (AW) and the wind tunnel consultant indicates that AW reverted back to **110 mph**.

FIGURE 2-2 Available documentation regarding wind speeds.
SOURCE: J. Abruzzo, L. Cao, and P. Ghisbain, 2022, "Arecibo Observatory: Stabilization Efforts and Forensic Investigation," Thornton Tomassetti, Inc., presentation to the committee, February 17, slide 34. Courtesy of Thornton Tomassetti, Inc.

---

[19] TT Final Report, Appendix J, p. 1.
[20] TT Final Report, Appendix J, p. 10.
[21] TT Final Report, p. iii (unnumbered in the report).
[22] TT Final Report, Appendix J, Figure 33, p. 35.
[23] TT Final Report, Appendix J, Figure 34, p. 35.

POST-MARIA ARECIBO TELESCOPE INSPECTIONS

There exists significant uncertainty about what Arecibo Telescope damage was evident immediately post-Hurricane Maria. The Arecibo Telescope standard "Preventative Maintenance Report" provided by the National Astronomy and Ionosphere Center's Arecibo Observatory (AO) for tower inspections comprises 10 "checkpoints" that are described and that are to be assessed by the observatory field staff.[24] These include the following: "Task 006: Note condition of cables socket and saddles. Verify that saddle does not show cracks, are [sic] completely painted, and are not rusted. If you find something of the above listed items, report to the maintenance office through your supervisor" and "Task 008: examine each cable carefully looking for evidence of broken wires."[25] The July 20, 2018, Tower 4 maintenance record, the earliest post-Maria inspection report available to the committee, reports "OK" for Task 006 and "No Broken Wires" for Task 008.[26]

Box 2-1 summarizes maintenance inspection on Tower 4 from July 20, 2018, to July 6, 2020. These maintenance records also noted no broken wires, and all systems were "OK."

Periodic inspections of the Arecibo Telescope's cables before Maria failed to note any progressive increases in the pullout of zinc from the cable sockets on multiple auxiliary and main cables during multiple inspections, including the M4N-T auxiliary cable. Figure 2-3 shows the typical strand pull-out observed on the M4S-T socket in 2003 (approximately ½ inch pullout, left image) and for comparison, the M4N-T pullout after Maria in 2019 (approximately 1⅜ inch pullout, right image). Note that the white paint is visible on the cable, and the extruded zinc is visible in the 2003 image. According to the TT Final Report,[27] the auxiliary cables were painted in 1995 and 2003. A comparison of the 2003 image with the picture from 2019 clearly shows significant additional pullout post-Maria. There are no images of this socket available from the inspections performed in 2018. The presence of a large area of unpainted zinc near the socket in the image from 2019 is a clear indication of major socket deterioration. The clear movement, or slip, of the cable at the zinc anchorages, no matter what the cause might be, should have been seen as a degradation mechanism in a key structural component whose failure would be catastrophic. It should have raised serious concerns about the condition of this and other cables, but there was no mention of such anomalies anywhere in the inspection reports.

As the zinc extrudes through the front of the socket, it is expected that a void or voids would appear somewhere within the socket (typically at the cap under the mastic coating) that can create a depression on the top of the socket. An example showing these depressions for sockets on the M4 tower end sockets is shown in Figure 2-4. This type of depression could have been noted by inspectors as an important clue as to the condition of zinc in the socket.

Additional significant increases in post-Maria socket pullout should have been considered a clear and unambiguous warning sign of a growing danger of catastrophic failure long before the failure of the auxiliary cable socket M4N-T.

Conclusion: The Arecibo Telescope operators would have benefited from more detailed engineering or structural risk guidance concerning inspection protocol, documentation, and/or indicators of structural deterioration and unexpected performance.

There was no documentation found by the committee of any immediate post-hurricane systematic inspections or loading analyses of the Arecibo Telescope's structural cables or socket connections beyond what was cited previously from the maintenance inspection reports. A structural analysis of the loads imposed on offshore structures subjected to a major (design level) hurricane event is required to demonstrate fit for purpose before putting it back into operation. TT observed,

---

[24] A. VanderLey, 2022, "Arecibo Observatory: Failure Event Sequence," National Science Foundation presentation to the committee January 25 (hereafter "NSF presentation"), slide 50.
[25] NSF presentation.
[26] NSF presentation.
[27] TT Final Report, Appendix D, p. 2.

> **BOX 2-1**
> **Preventative Maintenance Records: Tower 4 for July 20, 2018, to July 6, 2020**
>
> 1. July 20, 2018 (Monthly)
> 2. September 19–20, 2018 (Monthly)
> 3. October 24, 2018 (Monthly)
> 4. November 21, 2018 (Monthly) Noted Towers 4 and 12 inspected for the possibility of a robot to clean and paint the cables; robot could not pass past the dampers on the cables.
> 5. January 9, 2019 (Monthly) Noted all in normal conditions.
> 6. February 25, 2019 (Monthly) Noted all in normal conditions (subsequently more information available; see NESC/WJE report).
> 7. March 12, 2019 (Monthly) 4.8 mag earthquake in Salinas PR at 9:08 AM. Noted inspection of tower, stairs, saddle, cable sockets and cables. Noted all in normal conditions.
> 8. April 22, 2019 (Monthly) Tower inspected and safety lines measured for replacement. No other notes.
> 9. May 16–21, 2019 (Monthly) New safety line installation project.
> 10. July 31, 2019 (Monthly) Notes of maintenance activity unrelated to socket.
> 11. August 29, 2019 (Monthly) Routine inspection to all three towers due to Hurricane Dorian. No impact from Hurricane, no findings.
> 12. September 23–25, 2019 (Earthquake inspection) Two strong earthquakes noted at 6.0 and 5.1 mag at 11:35 am on 23 Sept 2019 were felt. Inspections to all three towers took place on 25 Sept 2019, no damages. On 24 Sep 2019, AO was in lockdown due to storm Karen. No damages from Karen.
> 13. September 27, 2019 (Monthly) Notes: Two strong earthquakes 6.0 and 5.1 @11:35am 9/23/2019 were felt. Inspections to all three towers took place on 9/25/19, no damages. A measurement scale was installed to the socket/cable Auxiliary 4 to monitor any movements. Inspected tower and noted that all the cables and the saddle were painted.
> 14. November 14, 2019 (Monthly)
> 15. December 28, 2019 (Monthly) Two strong earthquakes took place in Guanica (mag 4.8 at 6:00 pm and mag 5.1 at 9 pm). No damages noted.
> 16. January 6–27, 2020 Pictures noted to have been taken of the cable socket and saddles. No damage to the structure noted.
> 17. February 4–5, 2020 Another mag 5.0 earthquake in Guanica noted at 10:30 am. Inspections to all towers noted (This was the week a structural engineer was on site).
> 18. May 1–3, 2020 Inspections by drone due to COVID after 4.8 and 4.5 mag earthquakes were felt at AO.
> 19. June 4–5, 2020 Earthquakes noted (mag 4.4 and 4.6 mag); inspections were performed by drone.
> 20. July 3-6, 2020 Noted strong Earthquakes of mag 5.1; all observations were stopped. Inspections by platform personnel was [sic] performed. No damages [sic] were found."
>
> SOURCE: A. VanderLey, 2022, "Arecibo Observatory: Failure Event Sequence," National Science Foundation presentation to the committee, January 25, slides 51–52.

After Hurricane Maria in 2017, the Observatory staff observed and recorded cable slips of more than 1 inch on two of the sockets. There is no documentation to show whether these cable slips increased during the hurricane, and to our knowledge, they were not identified as an immediate structural integrity issue.[28]

The committee concurs and has found no evidence to the contrary. TT concluded that, just before the M4N-T failure, three of the six tower-end auxiliary cable sockets had a pullout of greater than ½ inch, and seven cables total had a pullout greater than ½ inch.[29] Even with these pullouts, TT was still assigning a safety factor of 2.0 or

---

[28] TT Final Report, p. ii.
[29] TT Final Report, Figure 25, p. 18.

FIGURE 2-3 Zinc pullout of auxiliary socket M4S-T in 2003 (left) and M4N-T in 2019 post-Maria (right).
SOURCES: G.J. Harrigan, A. Valinia, N. Trepal, P. Babuska, and V. Goyal, 2021, *Arecibo Observatory Auxiliary M4N Socket Termination Failure Investigation*, NASA/TM–20210017934, NESC-RP-20-01585, NASA Engineering and Safety Center, Langley Research Center, June 15, https://ntrs.nasa.gov/api/citations/20210017934/downloads/20210017934%20FINAL.pdf, Figure 6.4-2. *Left:* Photo from Ammann & Whitney, NAIC Cornell University, Arecibo Radio Telescope, Structural Condition Survey 2003, courtesy of Division of Rare and Manuscript Collections, Cornell University Library. *Right:* Photo from NAIC Arecibo Observatory, a facility of the National Science Foundation.

greater, which assumes zero strength degradation from creep or any other aging mechanism, to every socket with these largest cable extensions before the M4N failure.[30] This suggests that it was not recognized that material degradation modes such as creep can operate at a fraction of the cable's strength.[31]

The limited post-Maria cable socket pullout documentation is sparse. "Cable slips were observed but not consistently monitored and reported before August 2020."[32] The earliest known documentation of the post-Maria cable slip was a May 15, 2018, measurement of at least 1.5 inches of pullout at B12W-G, more than 6 months

---

[30] TT Final Report.

[31] Moreover, the connection between vulnerability to creep induced wire failure and pullout length is likely very complicated due to nuances in zinc seating, location of the crown of compression inside the socket and likely there cannot be assumed to be a direct proportionality between creep pullout as an absolute length and eventual socket failure.

[32] TT Final Report, p. 17.

FIGURE 2-4 Socket depressions on the M4 tower end.
SOURCE: Thornton Tomasetti, 2021, *Arecibo Telescope Collapse: Forensic Investigation Interim Report*, NN20209, prepared by J. Abruzzo and L. Cao, November 2; photo from NAIC Arecibo Observatory, a facility of the National Science Foundation.

after Hurricane Maria.[33] The last known cable slip documentation of socket M4N-T of (1.125 inches), where the first cable failure occurred, was observed on February 19, 2019,[34] more than 18 months after the hurricane but still more than 18 months before the failure. The observed slip of M4N-T was now at least one-third of the cable's diameter in 2019. "The inspection process did not couple the qualification/design process to a pass/fail criterion to trigger a replacement. Data were not provided to explain the decision process after the 2019 photos were taken."[35] The committee concurs with both these conclusions.

> Finding: Despite the dramatic pullout at multiple connections recorded by the AO field staff post-Maria, no additional concern about the evident structural distress was noted in the in-house or external inspections, and no action was undertaken to address it.

On February 18, 2019, photographs of auxiliary cable sockets on Towers 4, 8, and 12 documented the tower auxiliary socket pullout on all three towers.[36] Still, no inspection resulted in an emergency cease of operations due to concerns about the structural integrity of the facility. The reports from these inspections stated, "no structural damage noted." The structure clearly exhibited evidence of socket slip and wire breakage, but the structural damage in many cases was hidden and not assessable by visual observation. WJE noted, "In-service inspections showed evidence of progressive zinc extrusion on several Arecibo sockets, which in hindsight was evidence of cumulative damage and, in effect, a missed opportunity to prevent cable failure."[37]

> Finding: The routine inspection form used by the AO field staff did not include instructions for socket inspection.

---

[33] TT Final Report.
[34] TT Final Report.
[35] NESC Report, pp. 114–115.
[36] TT Final Report, Appendix D, pp. 14–18.
[37] WJE Report, p. 11.

TT eventually concluded,

> Cable slips were observed on the auxiliary sockets of the Telescope structure, and an increased amount of slip was observed on some auxiliary sockets within the past 2 years before the first cable failure. The cable slips were evidence of structural distress in the sockets [not governed by engineering safety factors related to cable strength divided by load] and should have raised a concern that cables may fail.[38]

NASA concluded, "In-service inspections showed evidence of progressive zinc extrusion on several Arecibo sockets, which in hindsight indicated significant cumulative damage."

Zinc pullout at a socket front meant that voids or depressions would necessarily appear either inside the socket (hidden voids) or on the socket back end (visible depressions). These depressions were never noted in any of the inspections, even though pictures show depressions forming on the back ends of several cable sockets (on both auxiliary and main cables) after the first cable failure. Cable pullout as a sign of growing socket distress was not recognized by either staff or contracted structural engineers.

Further cable pullout was not evident in the limited photographic evidence taken before and after the January 2020 earthquakes.[39] The M4N cable was predicted to be the second most lightly loaded auxiliary cable in the earthquake.[40]

> Initial report (January 27, 2020) noted damage to vibration dampers, tie-down blocks and slabs, potential for cracked platform steel components, and cracks in concrete buildings. Also, concerns were noted about site power infrastructure, the need for more safety and inspection equipment, and structural analysis and modeling for resiliency. [A] $3.325 million proposal submitted in 2020 and awarded included tasks for acquisition and installation of new vibration dampers to reduce vibration caused by external forces (wind or seismic events).[41]

The best available information on the distribution of the Arecibo Telescope's socket pullouts just before the M4N-T socket failure is summarized in Figure 2-5.[42]

Conclusion: The Arecibo Telescope's failure sequence began with Hurricane Maria.

THE HURRICANE MARIA AFTERMATH

The first evidence of additional cable socket slippage after the 2011 A&W survey was documented by Arecibo staff in 2019, more than 2 years after Hurricane Maria. The 3 or so years after Maria and before the Arecibo Telescope's collapse were occupied with repair planning, proposal writing, cost justifications, and administrative activity unrelated to the cables that failed.[43]

Ray Lugo of the University of Central Florida (UCF), and former director of UCF's Florida Space Institute and principal investigator for the National Science Foundation (NSF) operations grant of the AO, told the committee that he inspected the facility post-Maria as part of the handover to UCF: "I have used wire ropes and socketed cables lifting devices, and in my experience as an engineer, I'd never seen a cable, a socketed cable, with that amount of pull-through. I was told that it had been reviewed by an external engineer, and that there was not any real concern about cable failure."[44] NSF has reported that it does not have a record of this exchange. The consultants failed to notice or recognize the critical importance of the socket cable pullout or even the importance of immediately

---

[38] TT Final Report, p. 25.
[39] TT Final Report, Appendix D, pp. 16–17.
[40] TT Final Report, Appendix K, p. 9.
[41] NSF presentation, slide 20.
[42] TT presentation, slide 28.
[43] R. Lugo and F. Cordova, 2022, "Perspectives on Grant Award and Operations of Arecibo Observatory Cooperative Agreement by the University of Central Florida," University of Central Florida (UCF) presentation to the committee, February 17 (hereafter "UCF presentation"), meeting transcript minutes 00:09:58 to 00:21:54.
[44] UCF presentation, meeting transcript minute 00:08:46.

THE COLLAPSE: WHAT HAPPENED 27

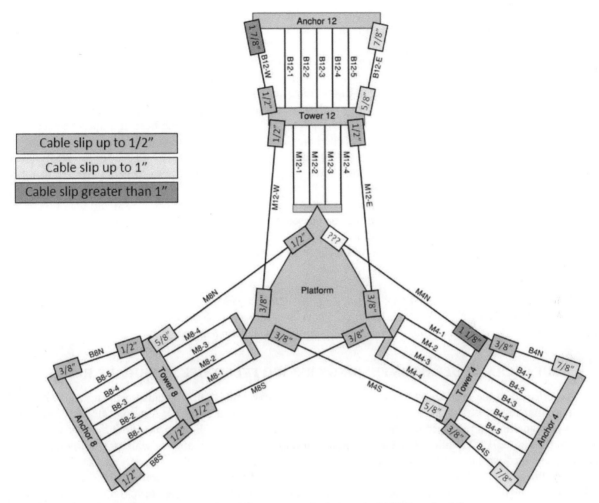

FIGURE 2-5 Arecibo Telescope socket pullout (slip), post-Maria, before the M4N-T socket failure.
SOURCE: J. Abruzzo, L. Cao, and P. Ghisbain, 2022, "Arecibo Observatory: Stabilization Efforts and Forensic Investigation," Thornton Tomassetti, Inc., presentation to the committee, February 17, slide 28; courtesy of Thornton Tomassetti, Inc.

inspecting for cable and socket connection damage post-Maria, except for wire breaks. NSF has told the committee that "at no time during the review of damage post-Maria was increasing socket slippage ever brought to NSF's attention as one of the issues to be considered for further inspection, analysis or in need of a repair."

Conclusion: The unexplained significant cable socket pullouts should have alerted the consultants that assumptions about remaining cable strength based on initial design safety factors were likely no longer appropriate.

The AO's working conditions in the last months of 2017, immediately post-Maria, and during the last months of SRI's facility management were difficult. During the preparation of proposals for the repair of the telescope, as well as the operational transition between SRI and UCF, Ray Lugo noted that power was not fully restored until sometime in December 2017.

We couldn't even access the platform because the catwalk had been significantly damaged, and that's the way to go up to the platform. Then we spent some time having to repair the cable car, which is the only other way of going up into the platform. At that time, getting materials and communicating to purchase anything was, was impossible. So, it really took several months [and] it was really more visual inspections of whatever could be accessed.[45]

We didn't even write the [repair] proposal until ... the June timeframe of 2018.[46]

The catwalk repair was part of the $2 million time-critical repairs immediately following Hurricane Maria.[47] It should be noted that the committee did not receive a presentation from the previous operator, SRI.

The M4N-T socket slip warning sign was finally photographed in 2019 but not heeded in the subsequent 1½ years it took for this tower connection to fail. As cited previously, this cable was not under suspicion or slated for replacement. Even after the M4N-T socket failed, Ray Lugo told the committee that he reached out to WSP Global, Inc., who told him after some "calculations" that "yes, we [the Arecibo Telescope] still had a positive factor of safety, less than two, but, I think it was like 1.7 or 1.8."[48] Since the committee was not afforded an opportunity to meet with representation from the firm, the committee does not know if engineers at WSP did not observe the socket pull-out warning signs or noted them but dismissed them as being unimportant. Ray Lugo also told the committee, "[AO] tasked Thornton Tomasetti to come up with a structural model that we could use to analyze the structure. And as a result, it actually did confirm what WSP gave us. While we were not at a 2.0 or a greater than 2.0 factor of safety, we were still greater than, I think, 1.7."

Conclusion: All of WSP's and TT's calculated safety factors appear to be incorrectly based on the nominal cable design strength without consideration given to intervening degradation mechanisms and damage mechanics, which are highly time-dependent and had unknown implications.

BUREAUCRATIC DELAYS IN FUNDING ARECIBO TELESCOPE HURRICANE REPAIRS

As reflected in the post-Maria timeline, there were various administrative events occurring at the Arecibo Telescope in the process of trying to repair the hurricane damage during a change of Arecibo Telescope operators. NSF was changing contract operators from an SRI-led consortium to a UCF-led consortium months after Maria hit in September 2017.[49] While the managing organization changed, "a majority of staff (including director of facility) remained the same at the Observatory."[50]

Unfortunately, the planned post-Maria repairs of the Arecibo Telescope's structural components were technically misdirected toward components and replacement of a main cable that ultimately never failed.[51] Thus, none of the specified repairs would have forestalled the Arecibo Telescope's collapse, even if completed in a timely manner. NSF awarded $2 million to complete the most critical, time-critical repairs after Hurricane Maria.[52] Cable replacement was not identified as "time-critical." A larger repair program, including the Tower 8 main cable replacement, were not identified "to be an issue."[53] A more detailed inspection and review of the Arecibo Telescope's structural condition immediately following Hurricane Maria might have identified the measurable socket degradation leading to critical repair or stabilization.

The repair process—planning, proposal preparation, justification for repair scope and costs, NSF's examination and funding, and the planned execution—for these non-urgent items remained incomplete by the time the telescope failed 38 months later.[54] Ray Lugo described to the committee how months of his time during 2018 were spent

---

[45] UCF presentation, meeting transcript minute 00:29:50.
[46] UCF presentation, meeting transcript minute 00:27:05.
[47] NSF presentation, slide 12.
[48] UCF presentation, meeting transcript minute 31:56.
[49] TT Final Report, p. 2.
[50] NSF presentation, slide 9.
[51] NSF presentation, slide 15.
[52] NSF presentation, slide 12.
[53] NSF presentation, slide 21.
[54] NSF presentation, slide 14.

writing, resubmitting, and justifying repair funding proposals.[55] Repairs had to go through the traditional "bid and proposal" process, described in more detail below, which added years of delay.[56]

The initial funding for the Arecibo Telescope's post-Maria most time-critical repairs was not awarded until nine months after Maria. The first hurricane repair award of $2 million was made to UCF in the summer of 2018.[57] The eventual congressional appropriation of $14.3 million for the AO included a large number of non-critical items "restoring AO to world-class scientific capabilities."[58] The most time-sensitive were further prioritized in the first $2 million award made to UCF on June 1, 2018.[59] The most time-critical repairs included cable replacement analysis and design, but not cable repair or replacement. Cable replacement was not identified as time-critical.[60] Almost another year later, in the Spring of 2019, NSF held a merit review panel for a proposal for the remaining $12.3 million repairs to restore the telescope, which included structural engineers to assess the main cable replacement.[61] NSF then required a detailed project execution plan and detailed plans for the cable replacement, to be put together by Louis Berger, as well as a request for proposal for the work. Next, NSF submitted a waiver to the Office of Management and Budget to permit $11.3 million of $14.3 million to be spent over 60 months instead of 24 months, which "would permit time for careful evaluation and design of repairs."[62] In the summer of 2019, almost 2 years after Maria, $12.3 million (to be spent on a timeline through fiscal year 2023) was awarded to UCF to complete 14 prioritized tasks in the short term, none of which involved Tower 4 cable repair or replacement (the tower where all the cable failures occurred), only the planned Tower 8 spliced main cable replacement. The Tower 8 main cable was the only cable-related repair noted in post-Maria evaluation and funding requests, as the project execution plan prioritized replacing the spliced cable as the highest risk. Structural analysis, main cable replacement design, and assistance in the cable replacement and construction administration was to be overseen by WSP (who acquired Louis Berger, who acquired A&W).[63]

"NSF was informed that the WSP structural engineers visited the site in February 2020 to work on the spliced main cable replacement, and they performed inspection of the towers, cables, and platform primary structural elements while there, but no additional damage was noted during those inspections."[64] NSF reported that it does not have any evidence that WSP looked specifically at socket "pullouts." NSF told the committee that its documentation simply states that WSP completed a "thorough structural analysis."[65] As part of its post-Maria repair proposal to Felipe Soberal, director, operations and maintenance of the AO, Michael Urbach of WSP (formerly Louis Berger) did propose to review "anchorage slip conditions,"[66] but it does not appear this proposal was accepted. The committee was not afforded an opportunity to question WSP.

Conclusion: Analysis performed by consultants in response to expressed concerns was incorrect. The consultants did not identify the key socket failure risk following Hurricane Maria, and none of the proposed repairs would have saved the Arecibo Telescope from collapse.

Finding: Only after the first socket failure did the consultants focus technical attention on the cable sockets, but even then, failed to consider the degradation mechanisms of the sockets. By the time a socket failed, there remained only a few months to respond.

---

[55] UCF presentation, meeting transcript minutes 19:34 to 24:09.
[56] NSF presentation, slide 13.
[57] NSF presentation, slide 12.
[58] NSF presentation, slide 10.
[59] NSF presentation, slide 12.
[60] NSF presentation.
[61] NSF presentation, slide 13.
[62] NSF presentation.
[63] NSF presentation, slide 15.
[64] NSF presentation, slide 21.
[65] NSF presentation, slide 21.
[66] Letter to Felipe O. Soberal from Michael Urbach of Louis Berger, March 4, 2019, p. 2.

SEQUENCE OF CABLE FAILURE EVENTS

First Cable Socket Failure: M4N-T

The M4N-T cable failure occurred when the M4N-T AUX socket pulled out on August 10, 2020. At the time of the M4N-T cable socket failure, it was reported that the "telescope was operating normally"[67] in the early morning under good conditions, dead calm, and no new warning indicators right up to the moment of failure. Without such a warning, and had this failure occurred during normal working hours, there could have been serious injuries and even loss of life during the collapse.

Right up to the time of the M4N-T cable socket failure, even with 1.125 inches of pullout, the cable's safety factor was incorrectly assessed to be ~2.2.[68] At the time of the M4N cable failure, it was not the most heavily loaded auxiliary cable.[69] As noted earlier, during Maria, the analysis predicted that M4N was one of the more lightly hurricane-loaded auxiliary cables, and M8N was the least loaded. The calculated cable safety factors gave no consideration to degradation mechanisms.

Although one cable stabilization had been planned as part of the hurricane repair appropriation, it was not for the M4N cable that ultimately pulled loose. Stabilization with a friction clamp for the B12W-G backstay cable on Tower 12 was set to begin almost 3 months later, on November 9, 2020.[70]

The M4N-T failure is illustrated in Figure 2-6.[71] Zinc creep in the less constrained central "core" of the wire "broom" allowed the cable load to be slowly transferred to the outer wires of the broom. The outer wires of the broom begin to fail in overload. Each outer wire failure transfers incrementally more load to the remaining outer wires. Eventually, all but 1 of the 56 wires that failed were in the outer three rows of the cable.[72] Ultimately, "forty-four of the 56 wire fracture morphologies were cup-cone fractures, nine were shear, and the remaining three were mixed-mode fractures, which included a progressive failure mechanism believed to be hydrogen assisted cracking, or HAC (one cup-cone/HAC and two shear/HAC)."[73] The broom's center pulled out of the socket, and the M4N-T cable socket failed.

Between the M4N failure on August 10, 2020, and the M4-4 failure on November 6, 2020, AO staff inspected the structure regularly.[74] Inspection photographs were taken periodically of cable pullout of Sockets M4S-T, M8N-T, M12E-T, M12W-T, B4N-T, B4S-T, B8N-T, B8S-T, B12E-T, and B12W-T.[75] Immediately after the failure, four wires on M4-1 broke.[76] The failed M4N-T socket was removed from Tower 4 on September 23, 2020. Emergency stabilization plans were approved by the end of September 2020.[77]

The emergency repair proposal for stabilization was submitted on October 19 and evaluated until October 23 by NSF, Sabal Engineering, and the U.S. Army Corps of Engineers.[78] Even after the M4N-T failure illustrated how the remaining strength of the Arecibo Telescope's cable spelter sockets had been vastly overestimated, no consideration was given to the potential strength degradation of the other cables. Because the cable's original design strength was used in the calculations, and there was no consideration given to degradation, the safety factor of the M4-4 and M4-2 cables was still assumed to be 1.6[79] after the telescope had been stowed. TT concluded "that M4N carried a tension of 600 kips when it failed."[80] But TT assumed M4N's strength was the original undegraded 1,314

---

[67] TT Final Report, p. 13.
[68] TT Final Report, p. 18.
[69] TT Final Report, Figure 40, p. 29.
[70] NSF, "Report on the Arecibo Observatory, Arecibo Puerto Rico Required by the Explanatory Statement Accompanying H.R. 133, Consolidated Appropriations Act, 2021," https://www.nsf.gov/news/reports/AreciboReportFINAL-Protected_508.pdf, accessed June 1, 2023, pp. 1–2.
[71] TT Final Report, Figure 25, p. 25.
[72] NESC Report, p. 26.
[73] NESC Report, p. 41.
[74] TT Final Report, Appendix E, p. 4.
[75] TT Final Report, Appendix D, pp. 15–24.
[76] TT Final Report, Appendix E.
[77] NSF presentation, slide 31.
[78] NSF presentation.
[79] TT Final Report, Table 4, p. 32.
[80] TT Final Report, Appendix G, p. 11.

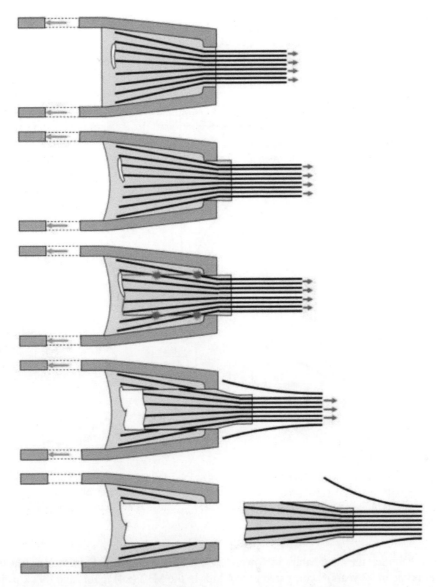

FIGURE 2-6 Schematic of wire broom geometry and failure sequence of socket M4N-T.
SOURCE: Thornton Tomasetti, 2022, *Arecibo Telescope Collapse: Forensic Investigation*, NN20209, prepared by J. Abruzzo, L. Cao, and P.E. Pierre Ghisbain, July 25, https://www.thorntontomasetti.com/sites/default/files/2022-08/TT-Arecibo-Forensic-Investigation-Report.pdf; courtesy of Thornton Tomasetti.

kips, giving it a safety factor of 2.2.[81] M4N-T's unanticipated failure illustrated that the safety factor calculated for the socket was inaccurate.

Even after the unanticipated failure of M4N at less than half its assumed strength rating, "up until November 6 [the] assessment, structural modeling, observations led to assessment that structure was stable enough for further stabilization work to proceed aided by additional monitoring devices, regular drone inspections, work safety plans, etc."[82] The assumptions made about the Arecibo Telescope's remaining cable strength after the M4N failure are summarized in Figure 2-7.[83] No consideration appears to have been given to any degradation mechanisms.

---

[81] TT Final Report, Appendix G, Figure 15, p. 12.
[82] NSF presentation, slide 33.
[83] TT Final Report, Appendix G, Figure 21, p. 16.

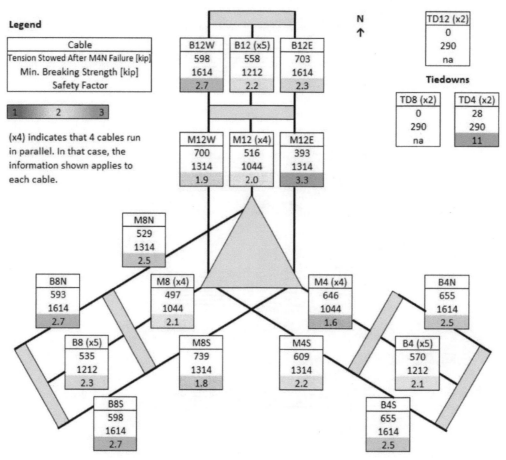

FIGURE 2-7 The TT calculated "Safety Factor" sockets post the M4N-T failure.
SOURCE: Thornton Tomasetti, 2022, *Arecibo Telescope Collapse: Forensic Investigation*, NN20209, prepared by J. Abruzzo, L. Cao, and P.E. Pierre Ghisbain, July 25, https://www.thorntontomasetti.com/sites/default/files/2022-08/TT-Arecibo-Forensic-Investigation-Report.pdf; courtesy of Thornton Tomasetti.

After the M4N-T failure, the main cables to Tower 4, M4 (×4) became loaded to 646 kips (the highest service load they had ever borne in 57 years)[84] compared to 516 kips in the M12 (×4) mains and 497 kips in the M8 (×4) mains, as illustrated in Figure 2-7. This reduced the TT computed (but incorrect) safety factor to 1.6, even with the telescope stowed because of the increased load.[85] However, the safety factor was actually near 1.0 because of a decreased cable strength at this load. This decrease in cable strength was due to hidden outer wire failures, which had already fractured due to shear stress from zinc pullout.

Second Cable Socket Failure: M4-4T

The tower end socket of M4-4 failed on November 6, 2020, and the failure sequence is illustrated in Figure 2-8.[86] TT's analysis calculated that there was 646 kips of load in this main cable at the time of failure, which had

---

[84] TT Final Report, Appendix C, Table 1, p. 3.
[85] TT Final Report, Appendix G, Figure 20, p. 16.
[86] TT Final Report, Figure 25, p. 25.

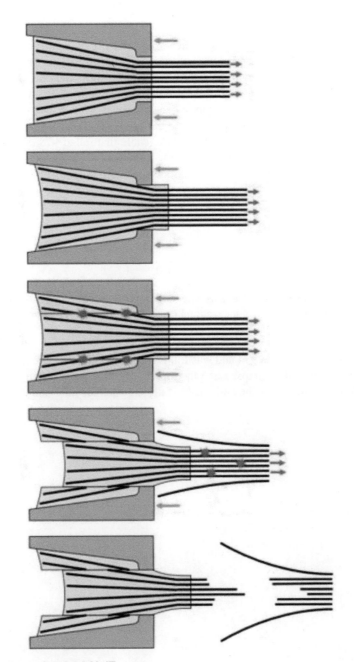

FIGURE 2-8 Failure sequence of socket M4-4T.
SOURCE: Thornton Tomasetti, 2022, *Arecibo Telescope Collapse: Forensic Investigation*, NN20209, prepared by J. Abruzzo, L. Cao, and P.E. Pierre Ghisbain, July 25, https://www.thorntontomasetti.com/sites/default/files/2022-08/TT-Arecibo-Forensic-Investigation-Report.pdf; courtesy of Thornton Tomasetti.

a nominally rated strength of 1,044 kips. The wire failed in the socket first, causing the core to displace, which led to overstress conditions in the cable wires outside the socket, causing total cable rupture.[87]

After the failures of the M4N-T and the M4-4T socket connections illustrated the remaining strength in the Arecibo Telescope's spelter sockets, the calculated value of the M4-2 cable safety factor was reported to be 1.3 in TT's 2021 *Arecibo Telescope Collapse: Forensic Investigation Interim Report*.[88] However, "this cable broke under conditions that should have been well within its support capabilities, indicating that it, along with the remaining main cables, may have been weaker than expected."[89] "[The] cable failed below its expected capacity, making it impossible for engineers to determine stability of structure (cable that failed was designed for 1,044 kips; was expected to hold 1,044 kips, but failed at 614 kips)."[90] TT, the Engineer of Record, advised NSF that another cable failure would be catastrophic. NSF determined that a "controlled decommissioning of the telescope" should begin.[91]

Final Cable Failure: M4-2T

The socket on the tower end of the M4-2 cable failed, and the Arecibo Telescope collapsed on December 1, 2020. According to TT,

> Between M4-4 Failure and Collapse (November 6, 2020, and December 1, 2020), the failure of M4-4 on November 6, 2020, caused multiple new wire breaks in the remaining M4 cables: three breaks at the tower end of M4-1, a break at the tower end on M4-2, and four breaks at the platform end of M4-2. No change was observed on the remaining tower sockets.[92]

M4-2 failed at its socket at the top of Tower 4 and was followed by the failure of the remaining two main cables. The suspended structure subsequently collapsed and the tops of all three towers then broke off. A drone "video shows a wire break on M4-2, followed by the entire cable failure 3 seconds later and the failure of the remaining two M4 cables immediately after."[93]

---

[87] TT Final Report, Appendix G, Figure 21, p. 16.
[88] TT, 2021, *Arecibo Telescope Collapse: Forensic Investigation Interim Report*, NN20209, prepared by J. Abruzzo and L. Cao, November 2 (hereafter "TT Interim Report"), p. 14.
[89] NSF presentation, slide 37.
[90] NSF, 2021, "Report on the Arecibo Observatory."
[91] NSF, 2021, "Report on the Arecibo Observatory," pp. 2–3.
[92] TT Interim Report, p. 25.
[93] TT Final Report, p. 20.

3

Analysis

CABLE SOCKET ZINC CREEP FAILURE

Thornton Tomasetti, Inc. (TT) did a more detailed analysis of six auxiliary cable sockets and concluded in its 2022 report (hereafter "TT Final Report"), "Four of the six sockets that were analyzed failed or were in the process of failing. The failures of sockets M4N-T and M4-4T occurred in the field and involved the rupture of multiple outer wires and a significant shift or complete pull-out of the core, which corresponds to core rupture."[1] An example of this failure process is generally illustrated in Figure 3-1. The specific processes in M4N-T and M4-4 will be discussed in more detail below.

Unfortunately, TT's socket analysis focuses only on sockets that had "failed or were in the process of failing."[2] Not all spelter sockets exhibited signs of creep, and not all were examined post-failure. Four of the five auxiliary main cable sockets attached to the platform exhibited only the ⅜ inch pullout.[3] TT's Lehigh University laboratory testing demonstrated that this pullout initially occurs during the initial "seating" of a newly poured socket when it first supports its service load.[4] The fifth platform axillary cable socket only had ½ inch of pullout, and no data were reported on the sixth socket. All the tower ends of the auxiliary main cables bore the same load (minus the 15.7 kips of cable weight) as their companion tower sockets. Yet five of six tower sockets had more cable pullout than their companion platform socket. While no data was available about the platform end of the sixth auxiliary cable, M4N, it would appear highly unlikely that the platform end of this cable had more pullout than the tower end's 1.125 inches[5] but went unnoticed and unreported. This pattern does not appear to have been noticed by TT, and none of the platform sockets that exhibited zero zinc creep in 23 years of service after 1997 were examined as part of the TT analysis to determine why they had no creep. It is also unlikely that each of the six platform sockets, manufactured at the same time and by the same people, had wire brooming superior to their six companion tower sockets. Thus, valuable insight into the effect of brooming was lost or why some sockets exhibited zero zinc creep.

---

[1] Thornton Tomasetti, Inc. (TT), 2022, *Arecibo Telescope Collapse: Forensic Investigation*, NN20209, prepared by J. Abruzzo, L. Cao, and P.E. Pierre Ghisbain, July 25, https://www.thorntontomasetti.com/sites/default/files/2022-08/TT-Arecibo-Forensic-Investigation-Report.pdf (hereafter "TT Final Report"), p. 44.

[2] J. Abruzzo, L. Cao, and P. Ghisbain, 2022, "Arecibo Observatory: Stabilization Efforts and Forensic Investigation," Thornton Tomasetti, Inc. (TT) presentation to the committee, February 17 (hereafter "TT presentation"), slide 28.

[3] TT presentation.

[4] TT Final Report, Appendix N, Figure 10, p. 7.

[5] TT presentation, slide 28.

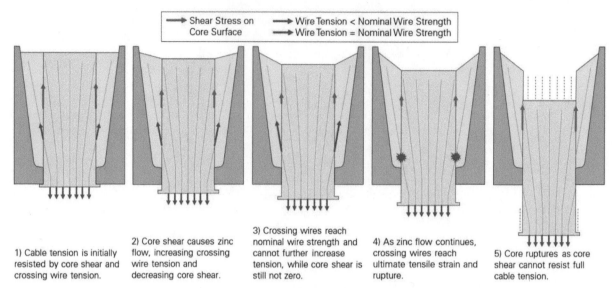

FIGURE 3-1 Core rupture when crossing wires cannot resist full cable tension and are fully developed.
SOURCE: Thornton Tomasetti, 2022, *Arecibo Telescope Collapse: Forensic Investigation*, NN20209, prepared by J. Abruzzo, L. Cao, and P.E. Pierre Ghisbain, July 25, https://www.thorntontomasetti.com/sites/default/files/2022-08/TT-Arecibo-Forensic-Investigation-Report.pdf; courtesy of Thornton Tomasetti.

Based on the observed Arecibo Telescope socket slip and the results of laboratory testing, it appears that all cable pullouts greater than ½ inch may involve zinc creep.

Conclusion: Based on the observed Arecibo Telescope socket slip and the results of laboratory testing, it appears that all cable pullouts greater than ½ inch may involve zinc creep.

Material creep is a time-dependent increase in permanent deformation in polycrystalline metals and alloys under loading at effective stresses that can be well below the material's yield strength.[6,7] Creep can also occur in compression, albeit at lower rates. Creep is very different from one-time work hardening at low temperatures brought about by rapid deformation. Work hardening requires additional higher stress to achieve additional permanent deformation once hardened. Creep deformation can involve softening over time or relaxation of the deformed state at a specific temperature, which then allows additional deformation at a fixed load. Creep deformation is a function of material properties, exposure temperature and time, and the applied load. A material's creep behavior cannot be quantified by short-term proof testing or loading to fracture quickly at high strain rates because there is insufficient time for softening. There is no substitute for long-term load exposure to quantify creep. Creep is extremely material dependent, and certain mechanisms have a very strong power law relationship to applied stress and temperature. Creep has three regimes: primary, secondary, and tertiary.

In primary creep, the strain rate decreases with time. In secondary creep, deformation continues at a steady state rate over long periods because recovery occurs dynamically such that work hardening in the zinc is balanced by recovery or softening effects. During long-term secondary creep at moderate stresses, significant permanent deformation can be achieved. The time-dependent dynamic recovery of the straining to accumulate additional permanent deformation (by dislocations and twinning) is balanced such that dislocation entanglements do not accumulate. The strain rate that can be maintained is relatively independent of total exposure time and a strong function of stress and temperature.

---

[6] A.J. Stavros, 1987, "Corrosion," *AMS Metals Handbook*, Ninth Edition, Volume 13, ASM International, p. 432.
[7] R. Abbaschian, L. Abbaschian, and R.E. Reed-Hill, eds., 2009, *Physical Metallurgy Principles*, Fourth Edition, Cengage Learning.

Creep follows a power law function of tensile stress but can even occur under compressive stress or pure shear stress. The creep strain rate depends on temperature, stress, shear modulus, vacancy diffusion rate, and grain size and is strongly materials dependent. Creep operates at stresses below yield and at temperatures typically above 40–50 percent of the absolute ratio of operating temperature to melting temperature (i.e., ~ T > 0.5 T_m), with temperature expressed in Kelvin. The range of $T_H = T(K)/T_m (K)$ is about 0.44 T_m at 85°F. Also, it should be noted that zinc creep can occur at a lower fraction of T_m. Furthermore, zinc creep is thermally activated, and there is no cut-off temperature since an Arrhenius-type behavior occurs. Zinc has a low activation energy for creep.[8] The possibility of zinc creep leading to failure would not have been predicted at the Arecibo Telescope, given the zinc deformation, expected service life, and design stresses.

Plotting the stress/shear modulus estimate and temperature ratio for pure zinc on the reported zinc deformation map[9] for 0.1 mm grains demonstrated that the Arecibo Telescope cable socket service lies in the regime of power law creep (PLC). The power law conclusion is also reported in the Wiss, Janney, Elstner Associates (WJE) report[10] and has been independently recalculated with the same findings of power law creep by the committee. It should be noted that power law creep is predicted when there is high stress, moderate temperature (T/T_m ~0.4, where T/T_m is known as also $T_H = T(K)/T_m(K)$, the homologous temperature [expressed in Kelvin]), and large grain sizes, all of which were operating conditions for the Arecibo Telescope spelter sockets. The dependency of creep rate on stress has been taken to the 4th power (assumed) by TT. The creep strain is estimated to be near 10^{-9}/sec, which also agrees with WJE's estimate. The effective "zone" of zinc that deforms by power law creep in a spelter socket is hard to estimate, as is the local state of stress in the complex geometry of the socket. However, if a 5 cm longitudinal zone of zinc near the spelter socket tension side aperture at uniform stress underwent power law creep, that could lead to deformations up to 1 cm over 25 years.

The finite element models presented in the NASA Engineering and Safety Center report[11] illustrate the detailed zinc behavior in the auxiliary cable's spelter socket. Figure 3-2 illustrates that the compressive stress pattern is a "spheroid" shape (red color).[12]

The compressive contact pressures form "arch actions" (compression struts) that transfer bond/shear forces developed between individual wires and zinc to the socket interior surface. The shear forces between the zinc and the wire are balanced by tension in the wires. Figure 3-2 shows various color bands (struts) representing different stress levels, with the red color signifying the arch with the highest compression. The arch stresses are highest near the mouth of the socket. These compressive forces (which correspond to transverse forces on the wires and the socket) produce shear (friction) resistance between the wire and the zinc in those areas. The compression stress at the socket is balanced by an equal and opposite force applied by the socket walls. The vertical components of these struts help resist the cable forces. The transverse arch action compression forces also contribute to the slanted (shear) wire failures observed on some outside wires in the M4N cable. Because of the higher stresses, the front of the socket is much more important than the back in resisting cable forces. This fact was also noted by Bradon and colleagues[13] in resin sockets.

Creep susceptibility depends strongly on the operating temperature relative to the material's absolute melting temperature. Since zinc has a low melting temperature, which is required to prevent the molten socket material from heat-treating the steel wires during socket filling, it is potentially susceptible to creep at tropical temperatures. Even when the socket material starts consistent with the ASTM standard, over time, there can be diffusion of elements or zinc dislocations to allow impurity-rich areas to develop that may change the properties of the parts.

---

[8] T.H. Courtney, 1990, *Mechanical Behavior of Materials*, McGraw-Hill.

[9] H.J. Frost and M.F. Ashby, 1982, *Deformation-Mechanism Maps: The Plasticity and Creep of Metals and Ceramics*, Pergamon Press.

[10] Wiss, Janney, Elstner Associates (WJE), 2021, *Auxiliary Main Cable Socket Failure Investigation*, WJE No. 2020.5191, June 21.

[11] G.J. Harrigan, A. Valinia, N. Trepal, P. Babuska, and V. Goyal, 2021, *Arecibo Observatory Auxiliary M4N Socket Termination Failure Investigation*, NASA/TM–20210017934, NESC-RP-20-01585, NASA Engineering and Safety Center, Langley Research Center, June, https://ntrs.nasa.gov/api/citations/20210017934/downloads/20210017934%20FINAL.pdf (hereafter "NESC Report"), Appendix C.

[12] NESC Report, Figure C-9, p. 560.

[13] J. Bradon, R.C. Chaplin, and I. Ridge, 2001, "Analysis of Resin Socket Termination for a Wire Rope," *Journal of Strain Analysis for Engineering Design* 36(1):71–88, https://doi.org/10.1243/0309324011512621.

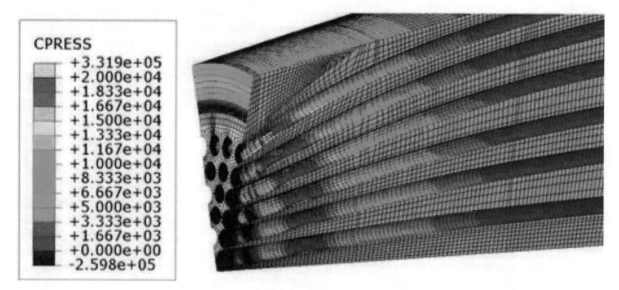

FIGURE 3-2 Finite element model of the auxiliary cable socket showing contact pressures (CPRESS) in pounds per square inch (psi).
SOURCE: G.J. Harrigan, A. Valinia, N. Trepal, P. Babuska, and V. Goyal, 2021, *Arecibo Observatory Auxiliary M4N Socket Termination Failure Investigation*, NASA/TM–20210017934, NESC-RP-20-01585, NASA Engineering and Safety Center, Langley Research Center, June 15, https://ntrs.nasa.gov/api/citations/20210017934/downloads/20210017934%20FINAL.pdf.

If excessive zinc creep occurs, it becomes increasingly difficult for the socket to maintain the compression forces on the wires associated with the arch action on the cable. Excessive creep produces a softening in the center of the socket. The wires at the center of the cable start to shed their mechanical load and redistribute the load from the cable's core wires to the outside wires. With further unloading of wires at the core, stresses on the outside wires increase until wire failure occurs in the most highly stressed wires. The examination of the failed M4N-T socket by NASA revealed multiple ductile (cup-and-cone) overloaded wire failures in the three outer wire rows around the perimeter of the cable near the front of the socket opening.

The structural modeling and analyses of wire brooming issues presented in the TT Final Report were based on some assumptions.[14] The compression stress helps the socket resist cable forces by transferring (through compression in zinc) the inclined arch action forces to the socket's conical surface. If pullout were to occur due to excessive creep of zinc, then the compression stress shown in Figure 3-2 (e.g., the red band) would also dissipate, and the compressive force at the surface of the core would dissipate significantly. Creep (or flow) of zinc in shear will also be accompanied by the creep of zinc in compression, thus dissipating the force in the struts. Furthermore, the angle of the compression stress was assumed to be constant (i.e., equal to the longitudinal slope of the interior surface of the socket). It was not adjusted as a function of the radius of the core. Figure 3-2 shows that the slope of the compression stress changes at different r values.

TT analyzed the sockets using the developed models to assess the effect of the extent and quality of wire brooming. They concluded that the effects of brooming imperfections were significant. However, it is important to note that the wire brooming imperfection significance is related to the extent of creep. When zinc creep becomes excessive, the compression stress weakens (due to the inelastic softening action in zinc), and the socket's load-

---

[14] The computational model presented in Appendix O of the TT Final Report estimated radial and shear stresses on a zinc core with a radius of r (with r ≤ the radius at the mouth of the socket). The shear stresses at the outside surface of the zinc core were calculated based on a friction equation (i.e., a zinc-zinc friction factor multiplied by the estimated normal force). It is not clear why the shear strength of zinc material was not calculated (in presence of a normal force) in lieu of the friction calculations. An assumption may have been made that cable pullout had already occurred (due to zinc creep or flow) causing separation of the core from the rest of the socket. If it is argued that friction should be used (instead of the shear strength of zinc at the surface of the core) because of the slip, then the normal pressure must also be reduced substantially.

ANALYSIS

carrying ability dissipates. In such cases, the effects of brooming issues become magnified and significant because of the loss of the arch action. Brooming imperfections without creep would, therefore, be less significant due to the beneficial contributions of a functioning set of compression stresses.

The NASA investigative report on the failed M4N-T cable socket pointed to "a socket design with insufficient design criteria that did not explicitly consider socket constituent stress margins or time-dependent damage mechanisms."[15]

> The socket attachment design was found to have an initially low structural margin, notably in the outer socket wires, which degraded primarily due to zinc creep effects that were activated by long-term sustained loading and exacerbated by cyclic loading.[16]

Nothing in the NASA conclusion points to anything unique about the Arecibo Telescope. The design margin for the Arecibo Telescope's sockets was typical of cable structures, and there was nothing unique about the Arecibo Telescope's cables or sockets. The NASA report does not explain why the common design characteristics of the Arecibo Telescope's sockets produced failure, which, to the committee's knowledge, had never been previously reported in similar zinc-filled spelter sockets.

> Finding: Core slippage and cable pull-out occurred by stage II creep more or less continually but slowly for over 23 years in the auxiliary cable tower sockets.

Partial pullout slippage by PLC would be expected to increase significantly when the zinc core pulls out sufficiently to diminish radial and longitudinal compression, as zinc is no longer as severely wedged in the socket. The committee believes this is a critical tipping point in the whole socket failure mechanism—that is, when core slip is advanced enough for stresses on zinc to switch from compressive wedge action such that radial compression enabled by the wedging action is relieved. As a result, the committee expects acceleration of PLC on the zinc core outside the socket aperture because it is strongly tensile stress–activated. Increased tension forces on the core are transferred to outer wires as indicated by TT and NASA, which ultimately can produce either slip or fracture. Cable tension and cable core slippage length were not directly tracible to cable pull-out probability because there are too many variables, and the sockets with the longest slip did not fail first. However, the stress dependencies are so strong in PLC that subtle differences in stress state produce large variations in creep rates observed.

Inspections performed on the Arecibo Telescope sockets in 2003 and 2011 by the structural engineering firm Ammann & Whitney (A&W) concluded, after inspecting the Arecibo Telescope auxiliary main cables in 2011, "As noted in the 2003 report, the cast zinc has separated away from the leading edge of all the sockets by up to 1/2."[17] Thus, it would appear there was no observed additional cable pullout in 8 years, so the expected pullout in the next 6 years, up to 2017, would most likely be small or non-existent.

As part of its forensic investigation, TT commissioned a socket load test, referenced earlier, conducted at Lehigh University's Fritz Laboratory. Socket B4S-G, a backstay socket, was recovered from the collapse along with 15 feet of cable and tested under static and cyclic loads. A new zinc spelter socket was cast at the free end of the cable. This added socket was required to complete the test but does not fully represent an aged spelter socket that might contain breaks. The specimen was installed in a test frame and loaded to 100, 125, and 150 percent of the cable's static load for 20 hours. Load cycles were applied at each of the load levels to measure the behavior of the socket under cyclic conditions. Finally, the cable was loaded to failure, which was slightly greater than the cable's minimum breaking strength.

Cable slip was monitored at each load level. The pre-existing cable slip before loading was measured at 0.875 inches on the B4S-G socket. Cable slip continually occurred at each of the loading levels for each socket. At the initial loading of 100 percent of the static service load, the additional cable slip in the new socket was 0.03 inches.

---

[15] NESC Report, p. 12.
[16] NESC Report, p. 12
[17] TT presentation, meeting transcript minute 02:33:36.

At the 150 percent loading, the measured displacement grew close to 0.1 inches over 24 hours.[18] The final displacement of the B4S-G socket was 1.375 inches (0.875-inch pre-existing slip +0.5-inch slip at failure). It should be remembered that the socket system is complex, and cable pullout alone could not reveal how many wires in the socket were already fractured during creep. "The slip rate increased substantially in B4S-G once the cable tension exceeded 75 percent of the Minimum Breaking Strength."[19] As the Lehigh University socket testing demonstrated, even a small amount of additional pullout would not be predicted to measurably impact socket strength.

Finding: The Lehigh University testing of an Arecibo Telescope socket with ⅞ inch of pullout confirmed that an Arecibo Telescope socket with only ½ inch of pullout would be expected to support the full strength of the cable with no hidden damage in the form of wire breaks.

Of course, some Arecibo Telescope socket strengths were clearly being degraded by accelerated zinc creep, a process that would have continued without Hurricane Maria. However, even though the cable loads from the wind of Maria were far below that which should have produced any additional cable pullout, this relatively small increase in cable loading appears to have significantly aggravated the creep rate of the ongoing cable pullout based on the measured cable pullouts from the A&W structural conditions from 2011 and post-Maria repair documents.

Conclusion: Based on available evidence, Hurricane Maria produced unexpected cable pullout and an unexpected acceleration in cable pullout.

Long-Term, Low-Current Electroplasticity

One of the open questions about the Arecibo Telescope's collapse is what unique circumstances of the Arecibo Telescope caused an unprecedented and significant acceleration in the spelter socket zinc creep. Searches by all the forensic investigators produced no previous reports of such a spelter socket failure mode despite more than a century of spelter socket use. To the best of the committee's knowledge, there are no previous reports of such spelter socket failure in more than a century of their use. The investigations did not produce a direct or plausible explanation of this unique phenomenon. The only hypothesis the committee developed that could possibly explain the measured patterns and ultimate effects of the observed socket zinc creep acceleration was the effect of electroplasticity (EP).

EP is "the reduction in flow stress of a material undergoing deformation on passing an electrical pulse through it."[20] EP was first discovered by Eugene S. Machlin, who reported in 1959 that making a 6 kV (dc) closed circuit with NaCl crystals made them weaker and more ductile.[21] Later, in 1963, Troitskii and Likhtman reported the same EP effect in zinc crystals.[22] In 2020, Baumgardner et al. reported the following:

A low-energy electroplastic effect in aluminum alloys at an ultra-low current density threshold between 0.035 and 0.1 A/mm^2 that resulted in EP-assisted reduction in hardness of 10% and increases in creep rate up to 38% over a range in temperatures from 25°C to 100°C. Systematic experiments and ab initio calculations showed that Mg-Zn alloying elements in Al7050 were the origin of the EP effect.[23]

---

[18] TT Final Report, Appendix N, Figure 10, p. 7.

[19] TT Final Report, pp. 39–40.

[20] A. Lahiri, P. Shanthraj, and F. Roters, 2019, "Understanding the Mechanisms of Electroplasticity from a Crystal Plasticity Perspective," *Modelling and Simulation in Materials Science and Engineering* 27(8), https://doi.org/ARTN 085006 10.1088/1361-651X/ab43fc.

[21] E.S. Machlin, 2004, "Applied Voltage and the Plastic Properties of 'Brittle' Rock Salt," *Journal of Applied Physics*, https://doi.org/10.1063/1.1776988.

[22] O. Troitskii and V. Likhtman, 1963, "The Effect of the Anisotropy of Electron and γ Radiation on the Deformation of Zinc Single Crystals in the Brittle State," *Doklady Akademii Nauk SSSR* 148:332–334.

[23] C.H. Bumgardner, B.P. Croom, N. Song, Y. Zhang, and X. Li, 2020, "Low Energy Electroplasticity in Aluminum Alloys," *Materials Science and Engineering: A* 798, https://doi.org/10.1016/j.msea.2020.140235.

"To date, the theories of Joule-heating, electron-wind force, and de-pinning from paramagnetic obstacles are the most commonly invoked explanations for instances of EP observed in different materials."[24] There is a more detailed and extensive listing of past EP research in Appendix C, along with references.

When an electrically conductive rod, cylinder, or cable is placed within an electromagnetic (EM) field, electrical currents are generated on the surface of the conductor at the same frequency as the EM wave. The resulting alternating electrical currents would be limited to a skin depth from the outside surface of the metal. In the Arecibo Telescope, the EM waves were high frequency (S-band radar at 2380 MHz), and thus, the skin depth would be small (on the order of microns). The Arecibo Telescope cables were not solid rods or cylinders and consisted of multiple layers of spirally wound round wires around a core element. This arrangement of wires complicates the current flow. Regardless, current always flows to a ground. The committee did not have access to the grounding circuits of the Arecibo Telescope, but, at a minimum, the towers would have had at least ground-connected lightning arrestors and extensive steel rebar from its top anchoring the saddle to the underground foundation. The lightning circuit and the rebar pattern were not described in any of the other reports, and no electrical measurements of the ground were made. It also appears from the pattern of cable pullout that the tower end of the auxiliary main cables may have had their ground connection path through the auxiliary backstay cables. In the Arecibo Telescope failure sequence, a relatively young auxiliary cable pulled out first at a load less than half its strength before two main cables that were more than twice its age pulled out of their sockets at much higher relative loads. This outcome could be explained by significant metal structural differences that affected the quality of the ground or the pattern of current flow (described below) through the zinc of their respective sockets at the tower end of the different cables.

All the Arecibo Telescope main and auxiliary cable wires terminated in (and were surrounded by) zinc within the cable end sockets. However, from an electrical perspective, the structure of the auxiliary backstay ground-end sockets and the auxiliary main tower-end sockets was substantially different at both ends from that of the original main tower-end sockets, as illustrated by Figures 3-3 and 3-4. The auxiliary backstay ground-end socket and the auxiliary main tower-end socket geometry were similar in that all the Arecibo Telescope's EM radiation–generated current had to flow entirely through the auxiliary cable socket zinc to reach ground. The electrical connection from the auxiliary main cable end socket to the auxiliary backstay cable end socket was provided by a metal box frame that surrounded the tower top, and to which both the auxiliary main and auxiliary backstay cables were attached, as shown in Figure 3-5. Consistent with these observations, the auxiliary backstay ground-end sockets and the auxiliary main tower-end sockets had the most significant measured cable pullout, as illustrated by Figure 3-6. This pattern was not discussed or explained in the other reports.

In contrast, the current generated in the main cables flowing to the tower end socket had a path to ground that only required current to flow through the front end of the socket zinc and may have also had alternate paths to the ground. The main cables on the tower end terminated in a socket whose front rested on the back of a metal saddle secured to the tower top. To reach their socket on the back side of the saddle, the main cables passed through saddle slots, which had enough clearance to permit the cable installation. Based on the Tower 8 cable replacement drawings, which do not call out a clearance value, there was so little clearance between the cable outside diameter and the sides of the saddle slot that the cable effectively "masked" the bottom of the slot from the white paint subsequently applied to the cable/saddle assembly, which could not get past the cable to cover the slot bottom. The main cable's paint masking of its saddle slot is shown in the drone frame capture after the failure of M4-4 shown in Figure 3-7. It also appears there could be some layers of torn lead paint that previously "bridged" the cable clearance with the saddle slot.

How much physical and electrical contact a main cable had with the saddle before its socket termination is unclear from the available evidence, and the resistance of these connections was not measured. If a main cable physically contacted any point of its saddle slot before its socket termination, this contact could provide the EM radiation–induced main cable current with an alternate parallel path to ground thus never reaching the socket zinc. Further evidence of physical contact of the main cables with their saddle slot can be seen in a photograph of the Tower 12 saddle, which is shown in Figure 3-8. Contact would appear to be the only mechanism to produce paint fracture below all four saddle slots. This same paint cracking below the main cable slots is also evident on the

---

[24] Lahiri et al., 2019, "Understanding the Mechanisms of Electroplasticity from a Crystal Plasticity Perspective."

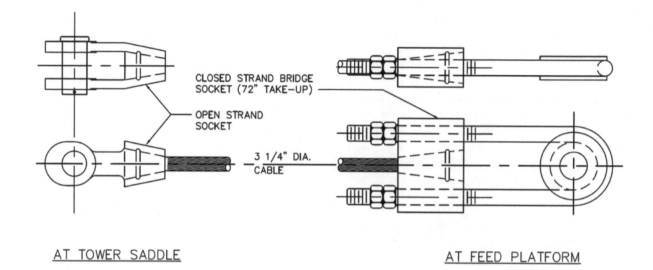

FIGURE 3-3 Auxiliary cable end socket geometry.
SOURCE: G.J. Harrigan, A. Valinia, N. Trepal, P. Babuska, and V. Goyal, 2021, *Arecibo Observatory Auxiliary M4N Socket Termination Failure Investigation*, NASA/TM–20210017934, NESC-RP-20-01585, NASA Engineering and Safety Center, Langley Research Center, June 15, https://ntrs.nasa.gov/api/citations/20210017934/downloads/20210017934%20FINAL.pdf.

Tower 4 saddle in Figure 3-7. However, the committee is unaware of any contact evidence in the form of fretting marks or similar contact indications on the cable wires.

The current generated on the outside surfaces of the auxiliary cable wires (and along the embedded length of the steel wires in sockets) would necessarily flow through all the zinc in the auxiliary main and auxiliary backstay sockets. All the current induced in the outside of the main cables did not have to flow through all the zinc in the socket, just through the front socket zinc near the front face of the socket bearing on the saddle. Main cable current may also have had parallel paths to ground of some quality in front of the socket through cable contact with the metal saddle, and "bridges" of lead paint (of unknown conductivity) that formed after the paint could not get through the small cable clearance. The auxiliary backstay cable pullout on the ground-end sockets of cables B12-W, B12-E, B4N, and B4S (Figure 3-6) was measured at ⅞ inch on three of the six auxiliary backstay cables and almost 2 inches on one of them, the most measured on any socket before failure. A potential explanation for this widespread ground-end cable pullout is that the auxiliary cable EM-generated current was flowing to ground through all the zinc in the ground-end sockets, which accelerated their creep.

Further evidence that the main cable-induced current did not all flow through their tower end socket zinc is the slower rate of main cable socket strength degradation from accelerated creep compared to the auxiliary cable sockets. The auxiliary cable tower socket M4N-T failed after just 23 years of service at a load ~46 percent (600/1314) of its nominal strength.[25] Despite being more than twice as old, with 57 years of service, the first main cable socket failure, of M4-4, only occurred after this socket exhibited ~62 percent (646/1044)[26] of its nominal strength for almost 3 months. Both Tower 4 sockets suffered significant strength degradation from accelerated creep, but no other explanation for this disparity in socket strength degradation rate has been offered.

Current flow, albeit at much higher levels than seen at the telescope, has been shown to soften and increase the creep of zinc in a much shorter time, as cited previously. But a lower current over a much longer time could also

---

[25] TT Final Report, Appendix G, Figure 15, p. 12.
[26] TT Final Report, Appendix G, Figure 21, p. 16.

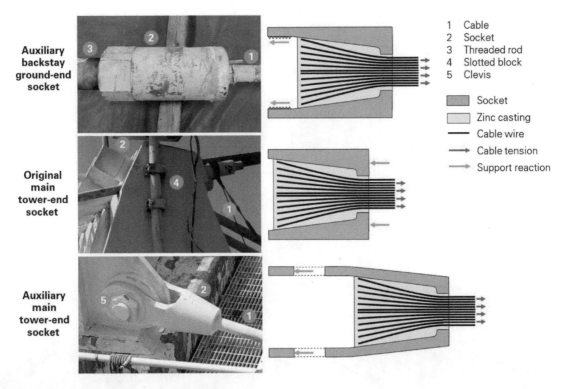

FIGURE 3-4 Spelter socket design used in telescope.
SOURCE: Thornton Tomasetti, 2022, *Arecibo Telescope Collapse: Forensic Investigation*, NN20209, prepared by J. Abruzzo, L. Cao, and P.E. Pierre Ghisbain, July 25, https://www.thorntontomasetti.com/sites/default/files/2022-08/TT-Arecibo-Forensic-Investigation-Report.pdf; with photos modified from SOCOTEC Engineering, Inc. (*top*) and NAIC Arecibo Observatory, a facility of the National Science Foundation (*middle and bottom*); courtesy of Thornton Tomasetti.

FIGURE 3-5 Tower cable termination for the main and auxiliary cables.
SOURCE: Adapted from drone footage from NAIC Arecibo Observatory, a facility of the National Science Foundation.

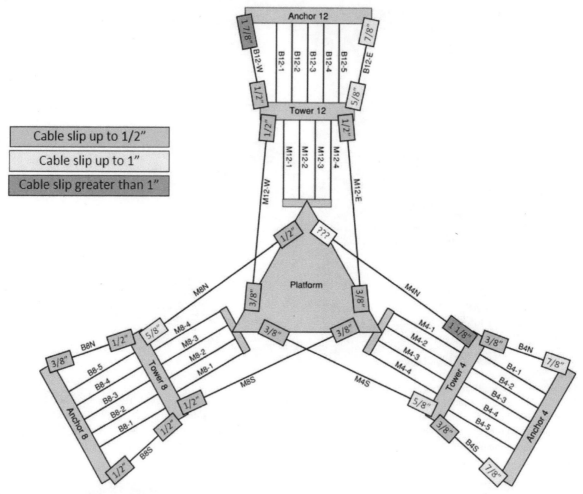

FIGURE 3-6 Post-Maria auxiliary main cable socket pullout before M4N-T failure.
SOURCE: J. Abruzzo, L. Cao, and P. Ghisbain, 2022, "Arecibo Observatory: Stabilization Efforts and Forensic Investigation," Thornton Tomassetti, Inc., presentation to the committee, February 17, slide 28; courtesy of Thornton Tomassetti, Inc.

plausibly increase zinc creep, but far less and much more slowly. No other mechanism for increasing the ambient temperature creep rate in high-quality spelter socket zinc at these cable loads has been suggested, discovered, or reported. Long-term slightly accelerated zinc creep would eventually compromise force transfer from the wires to the conical socket through zinc.

The relatively newer auxiliary cable socket that failed first, M4N-T, was not the most heavily loaded, nor was the brooming of this socket the worst found in the Arecibo Telescope's sockets. The reason(s) this auxiliary socket failed at less than half its strength—before 56 other main cable sockets that were more than twice its age—was not explained in any of the previously cited investigations. During a presentation of the TT Final Report, a representative of TT was not able to explain why this relatively young cable that had been determined to have a safety factor of greater than 2 suddenly failed.[27]

Considering that the long-term extrusion of zinc from the sockets and the subsequent pullout failure of the Arecibo Telescope cables have not been documented elsewhere, despite the long and wide use of zinc-filled cable

---

[27] TT presentation.

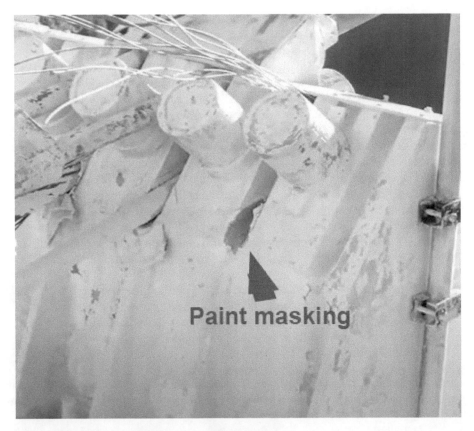

FIGURE 3-7 Drone picture of Tower 4 Saddle after cable M4-4 failure.
SOURCE: Adapted from National Science Foundation, 2020, "Video Footage of Collapse Arecibo Observatory," December 1, mark 0:56 minute, https://www.nsf.gov/news/special_reports/arecibo/arecibocollapseinfo.jsp; courtesy of NAIC Arecibo Observatory, a facility of the National Science Foundation.

FIGURE 3-8 Closeup of IMG_2173, cracked paint below all four main cable saddle slots on Tower 12.
SOURCE: Photo IMG_2173 from January 27, 2020, inspection of Tower 12 provided to staff, with labels added by the committee; courtesy of NAIC Arecibo Observatory, a facility of the National Science Foundation.

spelter sockets, an investigation of the Arecibo Telescope's collapse should explore possible site-specific factors that may explain the Arecibo Telescope's socket failures. Low-current electroplasticity (LEP) offers potential explanations for the questions: Why did the Arecibo spelter sockets fail? When did this failure mode appear unique to Arecibo? and Why did a relatively new auxiliary socket fail first, before any main cable socket that was more than twice as old? LEP also offers a potential explanation of why all six auxiliary cable tower sockets had more cable pullout than their respective platform sockets on the same cable and why (at least) four of these sockets exhibited zero additional pullout in their 23 years of service after installation.

The type, size, length, and fittings of the Arecibo Telescope's cables (whether the original cables constructed in the 1960s or the auxiliary cables installed in the 1990s) were not unusual and were catalog-selected items. The applied loads were mostly (but not all) dead loads, which were also not unusual considering that these types of cables are widely used across many applications in a wide range of industries. The 2.2 safety factor, discussed previously, ensured the applied dead loads never reached even half the cable strength, although the main cables operated at a lower, 1.98, safety factor for 30 years before the 1997 upgrade.[28] Cables on the Arecibo Telescope operated in a unique EM radiation environment compared to the typical zinc spelter socket terminated cable; specifically, the Arecibo Telescope cables were suspended across the beam of the "most powerful radio transmitter on Earth."[29] The Arecibo Telescope had two high-powered radar transmitters, one at 430 MHz and another at 2,380 MHz continuous wave.[30]

Thus, the main and auxiliary stay cables of the Arecibo Telescope (and their sockets) were within the path of the high-power electromagnetic waves emitted by these transmitters. The tower tops (including the cable sockets at the top of the towers) were also directly exposed to radio frequency (RF) radiation. A 2005 Arecibo Telescope RF safety report[31] states the following regarding maximum permissible exposure (MPE):

> RF field levels at the upper elevations of the three towers can exceed the MPE limits for Occupational/Controlled exposure under some conditions. The high RF fields occur when either of the feeds is tilted to a fairly high elevation angle and the beam of energy is aimed in the direction of the tower. Under these conditions, the upper elevations of the tower are within the main beam of energy from the antenna.

The RF safety report established that the average RF field strength at the platform or the tower tops, if they are illuminated, is above the RF MPE[32] limit. However, under other operating conditions, the larger illuminated area fields are not as strong. However, even under these circumstances, some current would be induced in the cables at an unknown strength. S-band radar from a source far less powerful than the Arecibo Telescope's has been found to induce skin current at a few hundred feet that can be sensed directly by people in the beam.[33]

Manufacturing methods based on EP employ very high electrical current densities, on the order of 10^3–10^6 A/cm^2,[34] to achieve the desired deformation quickly. The electrical current densities in the cables of the Arecibo Telescope would be expected to be orders of magnitude smaller but applied over orders of magnitude longer time (e.g., decades). The M4N auxiliary cable was ~23 years old when it failed, while the main cables were 57 years old, yet failed within months of each other. However, the main cable sockets became subject to a significantly higher relative load after the M4N failure, and their strength was less degraded. This disparity in longevity and degradation might be partially explained by the fact that both the main and auxiliary cables became subject to

---

[28] TT Final Report, Appendix C, Table 4, p. 9.

[29] A.P.V. Siemion, et al., 2011, "Developments in the Radio Search for Extraterrestrial Intelligence," *XXXth URSI General Assembly and Scientific Symposium*, https://doi.org/10.1109/URSIGASS.2011.6051263.

[30] J.L. Margot, A. H. Greenberg, P. Pinchuk, A. Shinde, et al., 2018, "A Search for Technosignatures from 14 Planetary Systems in the Field with the Green Bank Telescope at 1.15–1.73 GHz," *Astronomical Journal* 155(5), https://doi.org/10.3847/1538-3881/aabb03.

[31] RF Safety Solutions LLC, 2005, "RF Safety Report: An Analysis of RF Field Levels at the Arecibo Observatory," report prepared for Cornell University and the Arecibo Observatory, revised August 25.

[32] American National Standards Institute, ANSI Z136.1.

[33] B.E. Moen, O.J. Møllerløkken, N. Bull, G. Oftedal, and K.H. Mild, 2013, "Accidental Exposure to Electromagnetic Fields from the Radar of a Naval Ship: A Descriptive Study," *International Maritime Health* 64(4), https://doi.org/10.5603/IMH.2013.0001.

[34] H. Conrad, 1998, "Some Effects of an Electric Field on the Plastic Deformation of Metals and Ceramics," *Materials Research Innovations* 2(1):1–8, https://doi.org/10.1007/s100190050053.

ANALYSIS

the same most powerful EM current at the same time after the Arecibo Telescope's upgrade in 1997, but the zinc in the auxiliary cable sockets had to conduct all the induced current, whereas in the main cable, the sockets may have been just one of multiple parallel paths to ground.

EP may also explain both the symmetry and asymmetry observed in the auxiliary main cable socket cable pullouts, cited earlier, which are illustrated in Figure 3-6.[35]

At the time illustrated by Figure 3-6, 23 years after the 1997 upgrade was completed, four of the five possibly less well–electrically grounded (as explained above) auxiliary main cable platform sockets on cables M4S, M8S, M12-W, and M12-E show only their original ~⅜ inch cable pullout. This ⅜ inch of pullout occurred after the first 18 hours of full-service loading[36] of a newly made socket during the Lehigh University socket testing referenced earlier. There is no evidence that these four sockets experienced any zinc creep after they were put in service. This symmetry was not noted in the other reports.

All the tower end sockets of these auxiliary main cables, which experienced the same loading on the other end of the same cables, exhibited at least ½ inch of cable pullout (and some substantially more). Five of six tower socket cable pullouts on the auxiliary main cables exceed the platform socket pullouts on the same cable. While there was no date reported for the platform socket pullout of M4N, it is highly unlikely that a pullout greater than 1.125 inches on the tower end of this cable went unnoticed. This asymmetry was not noted either and is unlikely (odds of ~3 percent) due to random chance. It should be noted that the tower end of an auxiliary main cable has to bear the additional 15.7 kips of cable weight not borne by the platform socket. Still, this differential is substantially less than the cable-to-cable load variation or the load variation seen in the same cable as the azimuth arm rotated, as illustrated in Figure 3-9.[37]

The auxiliary main cable-to-cable load variation is more than 120 kips, and every auxiliary main cable sees a load variation of more than 40 kips. Yet, four of the five auxiliary man cable platform socket pullouts for which we have data were the same ⅜ inch of pullout from when they were manufactured and first loaded. It should be noted that the most lightly loaded cable in Figure 3-9 is M4N, whose tower socket failed first. Cable weight cannot explain the consistent asymmetry between the socket pullout of the platform and tower sockets of the same auxiliary cable.

Finally, differences in the interior cone shape between the platform and tower sockets do not appear to explain the asymmetric cable pullout. "Despite the exterior differences, the internal cable-to-socket connection was similar for all of the telescope's sockets."[38]

NASA concluded, "The unexpected vulnerability was further compounded by an effective design factor of safety that was significantly less than the minimum to ensure structural redundancy in the event of a cable failure."[39] However, it is important to distinguish factors of safety from redundancy. The Arecibo Telescope's collapse illustrates the inherent non-redundancy of the three-tower arrangement of the structure with the specified cable lengths. Had a fourth tower been present (at 90-degree angles), the chances for the swinging action and collapse would have been significantly reduced, and the structure would have had some redundancy.

Conclusion: A higher factor of safety would not have offered any additional structural redundancy to the Arecibo Telescope.

A lack of structural redundancy was built into the Arecibo Telescope's design in several places, such as the selection of three towers instead of four or the use of a single auxiliary cable instead of the original multiple parallel cables used for the main cables.

---

[35] TT presentation, slide 28.
[36] TT presentation.
[37] TT Final Report, Appendix H, Figure 13, p. 10.
[38] TT Final Report, p. 5.
[39] NESC Report, p. 12.

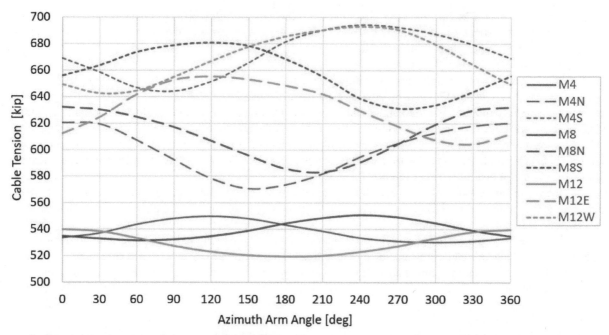

FIGURE 3-9 Cable load variation with azimuth arm rotation.
SOURCE: Thornton Tomasetti, 2022, *Arecibo Telescope Collapse: Forensic Investigation*, NN20209, prepared by J. Abruzzo, L. Cao, and P.E. Pierre Ghisbain, July 25, https://www.thorntontomasetti.com/sites/default/files/2022-08/TT-Arecibo-Forensic-Investigation-Report.pdf; courtesy of Thornton Tomasetti.

After the Aux M4N failure and before the main cable failure, additional analyses incorrectly asserted acceptable positive margin for the remaining structure despite no understanding of why a cable had failed at half the rated breaking strength. In hindsight, the structure was vulnerable to collapse after the Aux M4N failure.[40]

The committee concurs with this observation. Unfortunately, the profound safety implications of this realization were not noted. Retrospective observations about the impending failure warning from the observed cable slip found in the TT 2021 report, *Arecibo Telescope Collapse: Forensic Investigation Interim Report* (hereafter "TT Interim Report"),[41] are not found in the TT Final Report.

CABLE END SOCKETS

"The cable failures leading to the [Arecibo Telescope's] collapse occurred at cable ends, where cables are connected to supports with zinc-filled spelter sockets."[42] "Each failure involved both the rupture of some of the cable's wires and a deformation of the socket's zinc."[43] "Excessive cable slip occurs in zinc-filled spelter sockets due to zinc flow and is a sign of upcoming failure through core rupture or core flow-out."[44] All the analyses done by the investigators retained by NSF have arrived at the socket pull-out and/or rupture failure of spelter sockets by zinc "flow" as the root cause of the Arecibo Telescope's collapse. It should be noted that conventional zinc creep was a necessary but not sufficient time-dependent phenomenon to cause the Arecibo Telescope's collapse. The zinc

---

[40] NESC Report, p. 12.
[41] TT, 2021, *Arecibo Telescope Collapse: Forensic Investigation Interim Report*, NN20209, prepared by J. Abruzzo and L. Cao, November 2 (hereafter "TT Interim Report").
[42] TT Final Report, p. ii.
[43] TT Final Report, p. 1.
[44] TT Final Report, p. 49.

slip brought about a transfer of load to the cable's outer steel wires, which in turn failed by a combination of cup and cone ductile fracture, wire pullout, shear stresses, and, to a small degree, hydrogen environmentally assisted cracking (HEAC). The committee concurs with the TT Final Report that cable socket failure is the key structural element in the Arecibo Telescope's collapse.

Manufacturing the zinc spelter socket cable connections is a reasonably straightforward but manual process. The current practice requires extensive manual preparation of the cable end wires before they are individually broomed open, inserted into the socket, and encased in molten zinc, as illustrated earlier in the report (Figure 1-8). If not symmetrically broomed, some wires would be under higher tensile stress. The molten zinc is poured at 925–975°F (496–523°C). Zinc is used because its low melting point allows it to be poured over the cable wires at a temperature that will not cause the wire steel to appreciably weaken through undesired heat treatment. Conversely, overheating the zinc can affect its bonding properties and reduce the strength of the wires. All dross (foreign matter) must be removed from the surface of the pure molten zinc before pouring to prevent impurities from being poured into the socket.

Zinc is used for the sockets due to not only its low melting point but also its bonding to the galvanized surface of the wires, assisting their resistance to corrosion. Impurities of lead, cadmium, iron, and tin, which are products of the various extraction processes from sphalerite, zinc blend, or marmatite, must be controlled to prevent the formation of materials that are less resistant to corrosion. Three zinc samples were removed from the Arecibo Telescope's spelter sockets and tested. It is reported that "All three samples met the 99.5 pure zinc requirement of the ASTM B6 standard 14 prescribed for the original socket castings.[45] The NASA report concluded, "The Aux M4N socket build process and original construction was typical of zinc spelter open-end socket terminations."[46] The committee concurs.

Conclusion: The Arecibo Telescope's socket failures were not due to deficient materials or workmanship.

WIRE BREAKS

"As reported in Phoenix, Johnson, & McGuire, 1986, wire breaks in the main cables had been ongoing since early in the life of the telescope.... The records do not show any wire breaks in the auxiliary cable system."[47] This sentence appears true right up to the first cable failure. "Every known wire break is located near a socket at a cable end."[48] The first wire break appeared in the Arecibo Telescope's main cable M4-4 in 1962, before the telescope was even commissioned,[49] and the next wire break was reported in backstay cable B8-3 less than 1 year after the telescope's commissioning.[50]

What purported to be a comprehensive list of the Arecibo Telescope's wire breaks by location was presented by WJE, shown in Table 3-1.[51] TT also purported to present a comprehensive diagram of all the Arecibo Telescope's "known wire break locations and discovery dates before the first cable failure,"[52] which is shown in Figure 3-10. Both compilations reflect 40 wire breaks,[53] but unfortunately, these two compilations do not agree.

For example, starting with Tower 12, the WJE table reflects two subsequent breaks in the replaced B12-3 cable, whereas the TT diagram only shows one. Similarly, the WJE table shows a wire break on the M4-1 cable that is not reflected on the TT diagram. There is a significant disparity on the M4-4 cable. Both compilations show six breaks in this cable, but the TT compilation does not appear to reflect the 1962 wire break during construction.

---

[45] TT Final Report, p. 22.
[46] NESC Report, p. 25.
[47] Wiss, Janney, Elstner Associates, 2021, *Auxiliary Main Cable Socket Failure Investigation*, WJE No. 2020.5191, June 21 (hereafter "WJE Report"), p. 7.
[48] TT Final Report, Appendix D, p. 7.
[49] WJE Report, Table 2, p. 8.
[50] WJE Report, Table 2, p. 8
[51] WJE Report, Table 2, p. 8
[52] TT Final Report, Appendix D, Figure 13, p. 10.
[53] TT Final Report, Appendix D, p. 7.

TABLE 3-1 Main and Backstay Cable Wire Breaks Recorded by Wiss, Janney, Elstner Associates

Cable Number	Date	Location	Number
A12-3[a]	Dec. 12, 1966[b]	Lower (L)	2
	Sep. 6, 1974	L	3
	Jan. 22, 1976	L	1
A12-3 (1981)	Dec. 22, 1982	L	1
	Nov. 6, 2001	L	1
M12-2	Sep. 23, 1968	L	1
	Sep. 22, 1969	L	1
M12-3	Feb. 28, 1968	L	1
A4-1	Aug. 22, 1997	L	1
M4-1	Aug. 25, 1997	Upper (U)	1
M4-2	Dec. 5, 1969	L	1
	Aug. 26, 1970	L	1
M4-4	1962 Construction	L	1
	Jun. 22, 1967	L	1
	Jun. 30, 1975	U	1
	Jan. 17, 1983	L	1
	Jul. 28, 1983	U	1
	Nov. 23, 1988	U	1
A8-1	Mar. 20, 1970	L	1
A8-2	Jan. 7, 1971	L	1
A8-3	Feb. 11, 1964	U	1
A8-5	Jan. 27, 1964	L	1
	Dec. 18, 2001	L	1
	Jul. 9, 2003	L	1
M8-2	Nov. 20, 1967	U	1
M8-4	Mar. 20, 1973	L	1
	Jan. 28, 1983	L	1
	Nov. 6, 2001	L[c]	1
	Jan. 13, 2014[d]	L[c]	9

[a] Cable replaced in 1981.
[b] National Astronomy and Ionosphere Center, 2007, "Primary Suspension Schematic," DWG A-M02-005, February 23, https://naic.nrao.edu/arecibo/phil/hardware/telescope/140113_quake/Suspension%20Cable%20Breaks%20Diagram.pdf.
[c] Upper end of splice box.
[d] J.L. Stahmer, 2014, "Earthquake Damage to Main Support Cable M8-4," Ammann & Whitney, February 18.
SOURCE: Wiss, Janney, Elstner Associates, 2021, *Auxiliary Main Cable Socket Failure Investigation*, WJE No. 2020.5191, June 21; courtesy of Wiss, Janney, Elstner Associates, Inc.

ANALYSIS

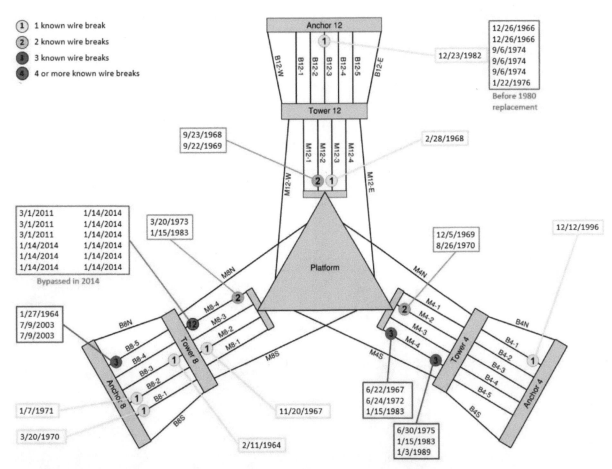

FIGURE 3-10 Known wire break locations and discovery dates before the first cable failure.
SOURCE: Thornton Tomasetti, 2022, *Arecibo Telescope Collapse: Forensic Investigation*, NN20209, prepared by J. Abruzzo, L. Cao, and P.E. Pierre Ghisbain, July 25, https://www.thorntontomasetti.com/sites/default/files/2022-08/TT-Arecibo-Forensic-Investigation-Report.pdf; courtesy of Thornton Tomasetti.

In contrast, the WJE table does not appear to reflect the second January 15, 1983, wire break at Tower 4. WJE reflects a November 23, 1988, wire break not on TT's diagram, and TT reflects a January 3, 1989, wire break not on WJE's table, both at the tower, so they are possibly the same wire break. Finally, in the Tower 8 backstay cable, both TT and WJE report six breaks, but some appear to be different wire breaks. TT is showing two wire breaks in cable B8-5 on July 9, 2003, and WJE is showing only one. In the Tower 8 main cables, TT reported three wire breaks in M8-4 on March 1, 2011, that do not appear on the WJE table.

The committee could not independently verify any of these wire break reports. A master compilation of these two sources of wire break data produces reports of 48 total wire breaks out of the estimated 5,256 wires in the original main and backstay cables. These breaks are expressed on a timeline in Figure 3-11. Reports reflecting only a few days reporting differences in the same location were treated as just one break.

After the 1975 upgrade, the overall rate at which wires broke was visibly reduced. This reduction was possibly related to a separate installation of a pressurized dry air system on each cable earlier in 1972.[54] Despite the dry sleeve installation, "Breakage occurred at roughly a constant rate from 1972 to 1978."[55] In 1981, the cable

---

[54] TT Final Report, Appendix D, p. 5.
[55] L. Phoenix, H.H. Johnson, and W. McGuire, 1986, "Condition of Steel Cable After Period of Service," *Journal of Structural Engineering* 112(6):1264, https://doi.org/10.1061/(ASCE)0733-9445(1986)112:6(1263).

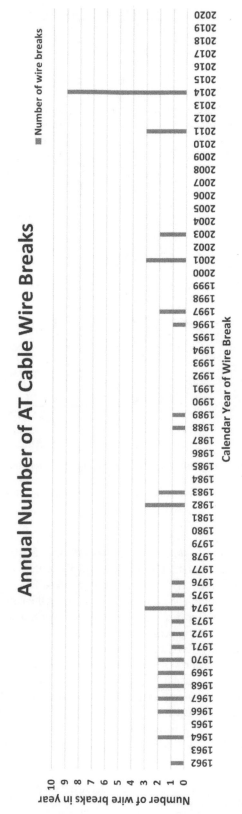

FIGURE 3-11 Annual number of Arecibo Telescope wire breaks by calendar year.

with the largest number of wire breaks, backstay cable A12-3, was removed and replaced, and the wire breaks were studied. "Analysts mentioned some possible evidence of both stress corrosion and corrosion fatigue, but the conditions of the fracture surfaces at the times of discovery were such that no definitive conclusions could be drawn."[56] It has been suggested that at least part of the cause of wire breaks could be the cable manufacturer removing the zinc galvanizing on the wires in the socket with acid before pouring the zinc. In doing this, they also removed some galvanizing for a short distance past the mouth of the socket. This gap in galvanization could lead to corrosion on the outer wires, resulting in breaks.[57] However, if removing the galvanizing was a significant contributor to wire breakage, the number of wire breaks would be expected to steadily increase with more time and corrosion, contrary to what was observed.

Another potential explanation for the reduced rate of wire breaks after 1997 is that the installation of the new auxiliary cables reduced the total force/stress in the main cables, but the structural analysis using the sag surveys in the TT Final Report suggests that the actual tension was consistent with the design and virtually the same on all three sets of main cables.[58] This explanation would imply the existence of some cable loading "threshold" for wire breakage, unrelated to time, that has not been reported.

The upgrade's powerful S-band radar may have facilitated LEP zinc creep by relaxing the zinc's shear grip on individual wires. Analysis of the combined wire break data reveals that a single cable, in a population of 27 cables, M8-4, is responsible for 15 of the 48 wire breaks. "The only new breaks on record after 2003 are located near the M8-4 cable splice."[59] If this single cable is removed from the data based on its unique circumstances, and the onset of the two different power levels of S-band radar is indicated, then Figure 3-12 illustrates a possible reason none of the auxiliary cables ever experienced a wire break. After the more powerful S-band radar came online in 1997, until the M4N-T failure, there were only four reported wire breaks in the following 23 years in the old main cables and none in the auxiliary cables, which operated their entire life in this more powerful S-band regime. It should be remembered that wire breaks only on the exterior of the cables could be seen and reported.

The fact that "Every known wire break is located near a socket at a cable end" could be the result of observational bias in that wire breaks in the suspended cable lengths would be more difficult to see from the towers or platform. As noted earlier, wire cable wire breaks were an explicit inspection item on the Arecibo Telescope's Preventive Maintenance Report.[60] However, even in the removed cable discussed previously, no mid-cable wire breaks were reported,[61] so this explanation is unlikely. The initial S-band radar was added to the Arecibo Telescope after 1974, with the completion of the first upgrade. The LEP-assisted zinc creep may have reduced the zinc's ability to tolerate the shear stress necessary to break a wire, or perhaps interfered with a corrosion mechanism. However, the potential impact of the LEP is confounded by the addition of cable moisture controls. The lower-frequency radars the Arecibo Telescope had when commissioned in 1963 do not appear to have impacted wire breakage. After 1975, the rate at which wires broke reduced visibly. The wire breakage appears to be reduced still further by the addition of more power S-band radar with the 1997 upgrade. The mechanism by which the radar could have reduced wire breakage is unknown, but possibly the more powerful S-band radar facilitated LEP zinc creep through some "relaxing" of the zinc's shear grip on individual wires to the point that wire breakage in the original main cable wires disappeared for 20 years. No wire ever visibly broke outside any auxiliary cable socket in the 23 years they operated, although it should be remembered that the auxiliary cables continuously operated at a higher factor of safety—that is, a lower relative load—than the main cables. Although wire breaks were found to be unrelated to the cable load or the likelihood of cable or socket failure, this virtual disappearance of wire breaks does not appear to have been noticed or explained.

Post the 1997 upgrade, the Arecibo Telescope maintenance records reflect documentation of "wire breaks," with all occurring in the main cables "near a socket at a cable end"[62] until 2003. No further main cable broken

---

[56] Phoenix et al., 1986, "Condition of Steel Cable After Period of Service," pp. 1264–1265.

[57] Phoenix et al., 1986, "Condition of Steel Cable After Period of Service."

[58] TT Final Report, Appendix G, Figure 15, p. 12.

[59] TT Final Report, Appendix D, p. 7.

[60] A. VanderLey, 2022, "Arecibo Observatory: Failure Event Sequence," National Science Foundation presentation to the committee January 25, slide 50.

[61] Op. cit., Phoenix et al., 1986, "Condition of Steel Cable After Period of Service," p. 1264.

[62] TT Final Report, Appendix D, p. 7.

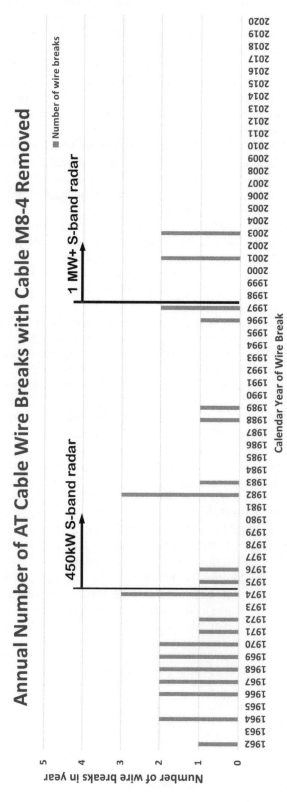

FIGURE 3-12 Arecibo Telescope annual wire breaks with cable M8-4 removed and S-band radar indicated.
NOTE: Dry air sleeves were installed on the cables in 1972 and on the auxiliary cables in 1995.

wires were reported right up to the first cable failure. No wire breaks were reported in any Arecibo Telescope auxiliary cable,[63] although clamps were observed at the platform end of the M12 cables, the tower end of the M8N cable, and both ends of the M8S cable.[64] The M4N cable that failed first had no reported clamps or wire breaks. Ray Lugo, University of Central Florida, told the committee that he asked the WSP, WJE, and TT team, "So what is the standard? You know, can you have five wire breaks? Can you have 10? Whatever? And so, I was actually told that there is no standard."[65] The number of wire breaks was found to be unrelated to the load in the cable.

> The M4N failure caused four new wire breaks at the tower end of in [sic] M4-4 Drone photos taken shortly after the failure show the deformation of the zinc casting of the M4-2 and M4-4 tower-end sockets. It is not clear whether these deformations occurred before or during the M4N failure. The zinc castings deformed in a pattern consistent with the cables slipping out of the sockets.[66]

However, the Tower 4 cable that suffered the most reduction in strength through wire failure, M4-1, with 11 known wire breaks,[67] never failed before the Arecibo Telescope collapse.

The aftermath of the second cable failure. "Shortly after the M4-4 failure, three new wire breaks were observed in M4-1, and a new wire break was observed in M4-2 near Tower 4 (Figure 16). Four wire breaks were also observed in M4-2 near the platform."[68]

Finding: Socket M4N-T had no broken wires and was the first to fail.[69] The main cable with the most wire breaks, M4-1 with 11 wire breaks, did not fail before the collapse. No correlation was found between the number of broken wires and the likelihood of socket failure.

Other processes that can cause failure at applied stresses below cable yield strength include corrosion, stress corrosion cracking, hydrogen-assisted cracking, creep, and fatigue. Each of these mechanisms was analyzed collectively in the WJE, NASA, and TT reports. Each mechanism was dismissed as the primary cause of failure at Arecibo based on the evidence at hand and analysis provided by TT, SOCOTEC Engineering, Inc., WJE, or NASA, except for zinc creep and "flow." A more detailed discussion of the Arecibo Telescope's potential failure mechanisms is found in Appendix B.

EARTHQUAKE

Puerto Rico experiences routine seismic activity because of its location along the boundary between the Caribbean and North American plates. "Since the telescope's completion in 1963, more than 200 earthquakes of moment greater than 4.5 occurred within 200 km (125 miles) of AO [Arecibo Observatory]."[70] On January 7, 2020, Puerto Rico experienced a magnitude 6.4 earthquake located on the southern coast of the island about 29 miles from the AO. This earthquake and the resulting aftershocks occurred after the auxiliary cable pullout was first noted and measured (May 2018 and February 2019) and 8 months before the first cable failure.

As part of its forensic analysis, TT performed a structural response analysis of the Arecibo Telescope for both the original design and upgraded structure. Recorded ground accelerations at the site (accounting for time delays based on the shear wave velocity of the ground near the surface at the site) were used as the boundary conditions for the analysis. As a comparison, TT also plotted the Design Earthquake Response Spectrum per ASCE 7-16

---

[63] TT, 2021, *Arecibo Telescope Collapse: Forensic Investigation Interim Report*, NN20209, prepared by J. Abruzzo and L. Cao, November 2 (hereafter "TT Interim Report"), Figure 7, p. 8.
[64] TT Final Report, Appendix D, Figure 15, p. 11.
[65] R. Lugo and F. Cordova, 2022, "Perspectives on Grant Award and Operations of Arecibo Observatory Cooperative Agreement by the University of Central Florida," University of Central Florida presentation to the committee, February 17, meeting minute 49:36.
[66] TT Interim Report, p. 11.
[67] TT Final Report, Appendix E, p. 14.
[68] TT Final Report, Appendix E, p. 9.
[69] TT presentation.
[70] TT Final Report, p. 12.

standard. The modeled earthquakes were "significantly less severe than the current design earthquake for the AO site."[71] In their finite element models, the maximum normalized stress range was 8 percent in the original structure and 5 percent in the upgraded structure. The lowest calculated safety factor for the earthquake loading condition before the first failure was 1.9.

WIND SPEED CONSIDERATION IN THE ARECIBO TELESCOPE'S DESIGN

The various engineering consultants, TT, WJE, A&W, and NASA, referred to different wind speeds in their reports. From Appendix J of the TT Final Report:[72]

The structural drawings for the original structure indicate a design wind speed of 140 mph. (p. 1)

Before the first upgrade of the telescope, a feasibility study by Ammann & Whitney (A&W) determined that the speed of 140 mph corresponds to the 300-year wind event at the telescope's site and that the 100-year wind speed is only 114 mph. A&W's structural drawings for the first upgrade indicate a design wind speed of 110 mph, suggesting that the 100-year wind event was selected as design event. (pp. 1–2)

A&W confirmed the two speeds of 110 mph and 123 mph for the global and local design of the upgrade respectively and, to our knowledge, the final design was based on those wind speeds. (p. 2)

First, the design wind speed was lowered from 140 mph to 110 mph between the original design and the second upgrade. (p. 2)

The use of "lowered" is misleading and seems to imply that the second upgrade's design used a lower design wind speed when these two different wind speeds are for different return periods, 300-year versus 100-year. These two wind speeds are essentially equivalent and should have resulted in the same member design when the appropriate factors and coefficients are applied.

GOVERNING CABLE DESIGN STANDARDS

The existing standard that governed the design and installation of structural cables in the Arecibo Telescope's upgrade was "Structural Applications of Steel Cables for Buildings."[73,74] However, the "Recommendations for Stay Cable Design, Testing, and Installation"[75] could have also been consulted. These two standards have fundamental and substantial differences in scope, design approach, qualification testing, inspection requirements, terminology, etc. The types of cables covered by the ASCE 19 standard are structural wire ropes and strands (except 7-wire prestressing strands). Strands are defined as "a plurality of wires helically twisted about an axis," and ropes are defined as "a plurality of strands twisted about an axis or about a core that may be a strand or another wire rope." The Post-Tensioning Institute (PTI) DC-45 standard addresses the design, testing, and installation of stay cables for cable-stayed bridges. The cables covered under this standard include those that consist of parallel wires, 7-wire strands, or bars. Stay cables used in modern cable-stayed bridges in the United States (since the 1990s) consist of multiple parallel greased-and-sheathed 7-wire strands terminated at anchorage plates using specially designed wedges. The parallel strands are then encased in high-density polyethylene pipes. The design of such stay cables is covered under the PTI DC-45 provisions.

Poured spelter socket cable end fittings (zinc or resin) and wire ropes/strands are not used in the main stay cables of modern cable-stayed highway bridges, but they are often used in pedestrian cable-supported bridges.

---

[71] TT Final Report, Appendix K.

[72] TT Final Report, Appendix J, pp. 1–2.

[73] ASCE (American Society of Civil Engineers), 2016, "Structural Applications of Steel Cables for Buildings," ASCE/SEI 19-16.

[74] This standard was produced by Committee 19 of the ASCE. The modern version of this standard was first published in 1996 (ASCE 19-96) and later updated in 2010 (ASCE 19-2010) and 2016 (ASCE 19-2016). The ASCE 19 standards traces back to an earlier standard published by the American Iron and Steel Institute (AISI) in 1966 (Tentative Criteria for Structural Applications of Steel Cables for Buildings).

[75] PTI, 2018, "PTI DC45.1-18: Recommendations for Stay Cable Design, Testing, and Installation," DC-45 Stay Cable Bridge Committee.

ANALYSIS

The design of the non-parallel wire ropes and strands is covered under the ASCE 19 standard. The cables are terminated at end fittings that may be poured sockets, swaged sockets, or mechanical loops with sleeve and thimble. These cables have been widely used over many decades in numerous industries (including naval, mining, construction machinery, and oil/gas). In bridges, the wire ropes/strands are still used in movable bridges and for hanger cables of major highway suspension bridges (but not the main cables of suspension bridges or stay cables in cable-stayed bridges). In the Arecibo Telescope, the type of cables used in the original 1960s design, as well as the cables added in the 1990s, were spiral wire strands with zinc-filled sockets, which would most suitably fall under the ASCE 19 standard.

While the PTI DC-45 standard has stringent and specific quality control and qualification test requirements (Section 4.0) that must be performed by independent testing laboratories, quality control, and cable testing under the ASCE 19 standard (when specified or left to the discretion of the engineer) can typically be performed and reported by the fabricator/manufacturer of the cable (e.g., Section 5.2). In PTI DC-45, the load requirements during the qualification tests are pre-defined and are not project-specific. All stay cables of a particular size, regardless of the calculated load demand in a particular location in the structure, must be subjected to the same loading (stress) condition meant to achieve a uniform level of high quality across the board.

Another major difference is the handling of the design requirements for the anchorage (in PTI DC-45 terminology) or end fittings (in the ASCE 19 terminology). In PTI DC-45, any failure in the anchorage components during qualification tests is cause for the rejection of the cable, implying that the strength in the anchorage during the final qualification test must be higher than the strength of the free length of the cable during the same test. In ASCE 19, end fittings are required to "develop an ultimate strength greater than the specified minimum breaking strength."[76] Therefore, under ASCE 19, the end fitting could have a lower actual strength than the cable itself as long as they have a higher strength than the "specified" (not actual) cable minimum breaking strength. In structural engineering, there is a strong tradition of not allowing the connection to control the design of the structure. The failure of the steel cable (in its free length away from the ends) would be associated with far more ductility than a sudden and brittle failure at the end connections. Therefore, the probability of failure at the connection (end fitting or anchorage) should be minimized to an acceptable level.

The cable design processes in ASCE 19 are currently based on the allowable stress design, while the design processes in PTI DC-45 are based on the load and resistance factor design (LRFD). A safety factor of 2.2 is specified in the ASCE 19 standard. In PTI DC-45, the load factors for various types of loads are varied based on the load combinations and requirements specified in the American Association of State Highway and Transportation Officials (AASHTO) LRFD Bridge Design Specifications.

The various consultants involved in this matter have referred to a standard published by AASHTO (AASHTO M 277) to assert that there is an acceptable "one-sixth of the diameter of the cable" limit for the pullout of the wires from the zinc sockets under service load conditions.[77] This pullout limit is a technically incorrect application of this standard.

The AASHTO M 277 defines the scope of the standard as follows:

> This specification covers steel wire rope for use in movable bridges. Both operating and counterweight ropes are included in nominal 6 × 19 rope construction. Suitable sockets are also included.

The AASHTO M 277 addresses a specific type of wire rope (6 × 19 construction) for a specific application (movable bridges). Section 7.5 of M 277 describes ultimate strength testing requirements and provides minimum strength values for 6 × 19 ropes up to 2½ inches diameter. The wire rope in the M4N-T socket was a vastly different 1 × 127 construction, which was 3.25 inches in diameter. Structural engineers contracted to evaluate the cable slippage and failures did not address this disparity between this standard's scope and the Arecibo Telescope's cables.

AASHTO M 277 Section 8.1 states:

---

[76] ASCE, "Structural Applications of Steel Cables for Buildings," Standard ASCE 19-16, Section 3.3.2, "End Fittings," p. 6.

[77] American Association of State Highway and Transportation Officials (AASHTO), 2019, "Standard Specification for Wire Rope and Sockets for Movable Bridges," M 277-06.

When stressed to 80 percent of its ultimate strength under the test specified in Section 7.5, to slip not more than one-sixth the nominal diameter of the rope. *If a greater movement shall occur, the method of attachment shall be changed until a satisfactory one is found.* [emphasis added].[78]

This important last sentence was not quoted in the various Arecibo Telescope forensic reports cited earlier.

Finding: The applicability of AASHTO M 277 to the very different Arecibo Telescope cables was never established by any consultant.

Finding: The specified cable loading at 80 percent ultimate strength for accepting a one-sixth diameter pullout was never attained in any socket of the Arecibo Telescope.

Conclusion: The one-sixth diameter pullout should not be an allowable limit of the Arecibo Telescope's cable pullout, especially considering the observed non-conforming condition of the sockets.

Before the failure of the first cable, eight Arecibo Telescope sockets, shown in Figure 3-13, exhibited pullouts of more than one-sixth cable diameter. Yet none of these cables were slated for replacement. Thus, it is not clear how the contracted structural engineers viewed the applicability of the AASHTO M 277 standard to the unique Arecibo Telescope configuration.

Finding: With a one-sixth diameter pullout measured at loads less than 50 percent of the cable's maximum strength (the loading of all the Arecibo Telescope sockets), the AASHTO M 277 criteria for permissible slip was not met.

According to Section C1.2 of the ASCE 19-16 standard, a proof load test may be performed on a cable assembly (typically 50 percent of the rated minimum breaking strength of the cable but not more than the pre-stretching force applied in the fabrication shop). The following statement is from Section C1.2 of the ASCE 19-16 (this statement was not present in the earlier editions of the standard):

Fittings attached with zinc spelter sockets will normally exhibit a small displacement of the zinc cone when seating into the socket during the proof test. This is observed where the cable exits the socket by comparing the positioning of the zinc at the socket base before and after the proof test. This displacement is a normal result of socket loading, and unless excessive, is not an indication of poor workmanship or design.[79]

ARECIBO TELESCOPE CABLE LOAD

The TT Final Report states that "excessive cable slips and eventual cable failures would not have occurred if the cable system had been designed with a safety factor of at least 3.0 under gravity loads" and recommended "using a safety factor of at least 4.0 under transient loads."[80]

The committee agrees with the NASA assessment[81] that the primary factor contributing to the failure of the M4N socket and the subsequent collapse of the structure was excessive zinc creep in the sockets. Long-term creep disrupted the force transfer mechanism in the sockets and redistributed forces from the strand's center wires to the outside wires as zinc pullout progressed. The fact that zinc creep was the primary factor is well established through microstructural assessments performed by NASA and the observed recrystallization of zinc.

However, the committee disagrees that a relatively high "gravity-load" fraction of the total load alone can explain the observed zinc creep and unprecedented failure of the Arecibo Telescope's sockets. TT presents no

---

[78] AASHTO, 2019, M 277-06, Section 8.1.
[79] ASCE, 2016, "Structural Applications of Steel Cables for Buildings (19-16)," Section C1.2, p. 36.
[80] TT Final Report, p. 50.
[81] NESC Report.

ANALYSIS 59

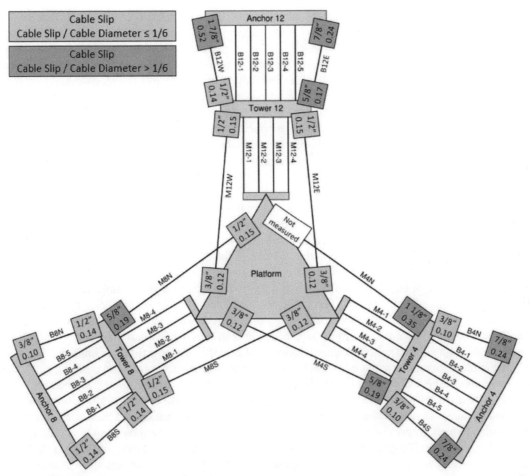

FIGURE 3-13 Cable slips on auxiliary sockets before the first cable failure.
SOURCE: Thornton Tomasetti, 2022, *Arecibo Telescope Collapse: Forensic Investigation*, NN20209, prepared by J. Abruzzo, L. Cao, and P.E. Pierre Ghisbain, July 25, https://www.thorntontomasetti.com/sites/default/files/2022-08/TT-Arecibo-Forensic-Investigation-Report.pdf; courtesy of Thornton Tomasetti.

evidence or data on the ratio of "gravity-load" to live load in any population spelter socket applications. Bridge literature generally indicates that a more significant live load requires a higher safety factor to address fatigue. The zinc-filled, spelter socket–terminated cable has a decades-long track record, is in widespread use across many industries, and remains a standard offering in many, if not all, cable manufacturer catalogs even today.[82,83,84] The committee has searched the literature and has found no other documented instances of this type of failure, although warnings to inspect spelter sockets for "any signs that the wires may be pulling out of the zinc" can be found in spelter socket manufacturer catalogs.[85] To the best of the committee's knowledge, there is no other documented failure of a zinc-filled cable-spelter socket via cable pullout in the past 100 years.

The head of the TT investigative team also noted the unprecedented nature of this type of failure.[86] Considering that this type of socket is ubiquitous in all types of industries, applications, and loadings, it is highly unlikely that

---

[82] Sullivan Wire Rope & Rigging, Inc., 2014, "Catalog," https://www.sullivanwirerope.com/catalog.
[83] Crosby, "The Crosby Group Catalog," https://www.thecrosbygroup.com/crosby-catalog, accessed June 1, 2023.
[84] Union, "Spelter Sockets," https://www.unionrope.com/products/slings/spelter-sockets, accessed June 1, 2023.
[85] Hanes Supply, Inc., "Technical Master Catalog," https://www.hanessupply.com/catalogs, accessed June 1, 2023.
[86] TT presentation.

the observed mode of failure is due to the influence of a higher dead load fraction of total load alone, and no such influence is found in the literature or cited by TT. Other site-specific factors must also be at play. The previous discussion of the uniquely powerful electromagnetic radiation of the Arecibo Telescope and the possible production of LEP deserves investigation since it remains the only site-specific factor the committee could uncover that might explain the substantial zinc creep acceleration and pullout of the Arecibo Telescope cables.

The NASA report also takes issue with the design safety margins for all the Arecibo Telescope's sockets, even those that have no evidence of any creep. In civil/structural engineering design, localized higher stress areas are not allowed to control the design of the entire component if the material is sufficiently ductile and can allow redistribution of stresses elsewhere in case of yielding. This approach has resulted in the successful long-term performance of these types of sockets. It is the uniquely excessive creep that made the difference in the Arecibo Telescope cables.

The engineering consultants evaluating the condition of the Arecibo Telescope structure (before and after the first cable failure) had consistently and extensively relied on their safety factor calculations based on an estimated ratio of nominal cable strength to cable load. While the cable tension can be estimated through detailed cable sag, the consultants' use of the original and unimpaired nominal cable strength (as capacity) was fundamentally unsound, especially considering the observed zinc pullout of the M4N-T socket at a load of less than half the cable's nominal strength and the subsequent failure of the first cable. Even after the M4N-T failure illustrated this discrepancy, the consultants continued calculating safety factors and assessed remaining strength based on the cables' original nominal strength. In a report written after the failure of M4N-T, one consultant wrote that "the current capacity/demand ratios for the primary structural elements and the suspension system as a whole are significantly greater than one rather than just barely greater than one" and "when there are no significant loads other than gravity acting on the system, failure of additional cables in the near future is unlikely."[87]

TT's recommendation to raise the safety factor in all spelter socket cable structural design from 2.2 to 3.0 or 4.0 means recommending a significant overdesign of the entire cable suspension system, driven by TT's design uncertainties of the capacity of the end connection sockets. The committee considers this recommendation to be unjustified by the evidence and based on TT's misunderstanding of what caused the unprecedented Arecibo Telescope socket failures. There is a long-standing tradition in civil/structural engineering of avoiding making the connections the controlling design elements, as connection failures are typically brittle and sudden. A change in the factor of safety from 2.2 to 3.0 or 4.0 in a design means that the number of wires or strands in a cable, that may be hundreds of feet long, must be increased by 36 percent and 82 percent, respectively, just to meet TT's uncertainty over the long-term performance of the cable end connection that has only manifested itself in a single unique installation in the past 120 years. Also, more steel for a given tension leads to higher cable sag, which increases local demand on anchorages for variable loading. If the issue of excessive creep were proven to apply to other cables (or had been observed in other cables) under high dead loads (i.e., not specific to the Arecibo Telescope's site conditions), then a reasonable approach would be to avoid using zinc sockets for such cables and maintain the existing factors of safety and the efficient quantity of steel used. Zinc sockets are not the only options for such major structural cables. None of the inclined stay cables used in major cable-stayed highway bridges built in the United States in at least the past 25–30 years includes any zinc-filled sockets similar to the Arecibo Telescope's cables.

Conclusion: The TT recommendation for raising the safety factor in ALL spelter socket cable structural design from 2.2 to 3.0 or 4.0 would result in significant increase in strand number and cable size. Connections should not be the controlling design elements.

Finding: The TT recommendation to increase the safety factor for socket cable design is not justified by the evidence. It is unclear how high the safety factor must become to sufficiently suppress power law creep in circumstances such as EM-induced LEP and high local stresses at specific locations in the Arecibo Telescope spelter sockets.

---

[87] UCF, 2020, "NSF Proposal Number: 2102922. Management and Operations of the Arecibo Observatory," proposal to NSF, October 19, page 86/495.

RISK CONSIDERATIONS

Probability of Failure

The current standard of practice for conventional structures produces an expected probability of failure on the order of 10^{-5} per year for components like cable M4-N (e.g., ASCE 7-22[88]). Based on this annual failure probability, the theoretical probability that this cable would fail in 25 years of service is small, much less than 1 percent. When the total number of cables and the service life is considered (27 original cables installed in 1963 and 12 auxiliary cables installed in 1997), the probability of at least one cable failing by 2020 is calculated to be about 2 percent.[89] Therefore, the failure of M4-N in 2020 is unexpected but possible based on this perspective from the standard of practice for conventional structures. However, it is questionable whether the standard of practice captured in ASCE 7-22 for structures primarily associated with steel and concrete frames applies to this telescope structure with a system of cables supporting a platform.

A structure somewhat analogous to the Arecibo Telescope is a floating offshore platform for producing oil and gas. The floating platform has a system of nominally 10 mooring lines[90] generally made of steel wire rope and subjected to relatively large, sustained tension loads that hold the floating platform on station. The value of this analogy is that there is experience available for hundreds of floating platforms with decades in service. An industry survey of mooring line failures over 13 years found that the frequency of failure for a single line was 2.5×10^{-3} per year.[91] If this failure probability is applied to the lines in the telescope, the probability of at least one line failing between 1963 and 2020 is 99 percent, and the expected number of line failures is 4.5 lines. This expectation of four to five failures of lines is consistent with the observed performance of the Arecibo Telescope's support system: one backstay cable was replaced in 1981 due to six broken wires; a cable connection was bypassed in 1997 due to damage during installation; a cable was bypassed in 2014 due to wire breaks in response to damage from a magnitude 6.4 earthquake; and cable M4N failed in 2020.

Furthermore, the frequency of multiple line failures in an event for offshore mooring systems was about 10 percent of the frequency for single line failures. Applying this analogous conditional frequency to the Arecibo Telescope support system gives a 40 percent conditional probability that at least two lines would fail together between 1963 and 2020.[92] Therefore, the failure of cables M4-N and M4-4 within several months of one another at AO in 2020 would not be unexpected, given the experience with offshore mooring systems.

The experience from offshore mooring system cables is also insightful into the patterns of line failures:[93] (1) the majority of failures occurred at or near terminations (connections); (2) failures were nearly always caused by a reduction in capacity rather than an overload; and (3) failures were more frequent both early in the service life (e.g., damage caused by handling and installation) and later in the service life (e.g., corrosion, fatigue, stress corrosion cracking [SCC], or HEAC). A notable conclusion from this experience was that using a larger factor of safety in design would generally not have prevented failures but may have delayed them.[94] The committee could find no evidence that any mooring line experience was considered in the Arecibo Telescope's original design or

---

[88] ASCE, 2022, "Minimum Design Loads and Associated Criteria for Buildings and Other Structures," ASCE 7-22, Table 1.3.2, for a structure in Risk Category I.

[89] This calculation assumes that the chance of failure is independent between cables, which is reasonable since the governing uncertainty is the capacity of the cable and factors that may affect it from fabrication, handling, installation, and degradation processes over time in service, and that the chance of failure is a constant with time, which is debatable since there can be degradation with time. However, the target probability of failure given in ASCE 7-22 is a nominal value for a 50-year service life that implicitly accounts for the possibility of degradation with time.

[90] The word line is used here to include the cable and connections.

[91] E. Fontaine, A. Kilner, C. Carra, D. Washington, et al., 2014, "Industry Survey of Past Failures, Pre-Emptive Replacements and Reported Degradations for Mooring Systems of Floating Production Units," OTC 25273, Houston, TX.

[92] The probability of failure per line is assumed to be 2.5×10^{-3} per year; the probability of failure per year for at least one line in the system is $1 - (1 - 2.5 \times 10^{-3})^N$ where N is the number of lines in the system during that year; and the probability of failure per year for multiple lines in the system is assumed to be 10 percent of the probability of failure for at least one line in the system.

[93] P. Smedley and D. Petruska, 2014, "Comparison of Design Requirements and Failure Rates for Mooring Systems," Proceedings of the Offshore Structural Reliability Conference, American Petroleum Industry, Houston, TX.

[94] NESC Report.

its subsequent upgrade. The committee also found no evidence that this experience was consulted after any cable failures or the subsequent analysis of the Arecibo Telescope's collapse by any of the involved experts.

Consequences of Failure

The consequences of even one cable failure for the Arecibo Telescope system were significant. The failure of line M4N-T in August 2020 caused damage to the dish and the Gregorian dome on the access platform as the line swung away from the tower, making the telescope inoperable. When one end of a 700-foot cable drops, with one end still supported at a 500-foot elevation, the unsupported 200 feet of cable that smashes into the ground weighs approximately 2 tons. This initial line failure could have injured or killed workers, and even visitors if it had happened later in the day rather than in the middle of the night. The National Science Foundation (NSF) established a safety zone in the vicinity of the dish after this unexpected cable failure.[95] A week after this first cable failure, "NSF requests safety plan prior to approving work at site" and "communicates to UCF [University of Central Florida] and AO" that "safety of personnel is the highest priority."[96] The failure of a second tower cable in November 2020 was followed by the complete Arecibo Telescope collapse 3 weeks later. The complete collapse destroyed the dish, the access platform, and the Gregorian dome; destroyed the support towers; and damaged the Visitors Center and the Learning Center. This complete collapse could have injured or killed workers if it happened a few weeks earlier.

There are, again, useful comparisons concerning consequences between the telescope support system and mooring systems for offshore platforms. As with the telescope, offshore mooring systems are generally designed so that the loss of at least two lines is necessary to lead to a failure of the mooring system, which means a failure to keep the offshore platform on station. However, in contrast to the telescope, failure of an offshore mooring system is generally not catastrophic because there are safety systems to minimize hydrocarbon releases if the platform moves off station, and the platform itself does not typically sink (although it may cause collateral damage if it impacts other vessels off station), a broken line is not likely to hit personnel, and offshore platforms are evacuated in advance of hurricanes.

Risk Management

The risks associated with failure in a structural support system can be managed by reducing the probability of a failure or reducing the consequences of failure. For the Arecibo Telescope, several measures were taken to reduce the probability of failure. The original cable support system had a redundant cable design for the three towers, with four main cables along each load path. Corrosion protection was applied and maintained throughout the service life. The cables were regularly inspected for broken wires, and cables were periodically replaced or bypassed when showing signs of distress. The system was inspected after major events, including hurricanes and earthquakes.

However, there were several notable measures not taken that could have further reduced the probability of the Arecibo Telescope's support system failure. First, the auxiliary cable system to support weight added to the platform in 1997 was designed with only one cable along each load path; the platform shifted and rotated suddenly when cable M4N-T failed, distributing loads in an uncontrolled (not-designed-for) manner to the remaining cables. Second, action was not taken quickly when a cable needed repair. Only the damaged M8-4 cable was scheduled for repair after Hurricane Maria, but it still had not been repaired/replaced by the time cable M4N failed 3 years later.[97] When M4N failed, there was no plan in place to quickly replace it to restore the system before M4-4 failed 3 months later. Third, the service life of the system was not clearly defined after the 1997 upgrade. The longer a structure is in service, particularly in a corrosive environment with loading from hurricanes and earthquakes, the more likely it is that damage will occur. Fourth, a detailed risk analysis was not conducted.

---

[95] NSF, "Report on the Arecibo Observatory, Arecibo Puerto Rico Required by the Explanatory Statement Accompanying H.R. 133, Consolidated Appropriations Act, 2021," https://www.nsf.gov/news/reports/AreciboReportFINAL-Protected_508.pdf, accessed June 1, 2023, p. 1.
[96] A. VanderLey, 2022, "Arecibo Observatory: Failure Event Sequence," National Science Foundation presentation to the committee January 25 (hereafter "NSF presentation"), slide 28.
[97] NSF presentation, slide 15.

Often, a failure modes and effect analysis (FMEA) is sufficient. In this case, the value of the functional system to the science program, coupled with its novelty and a lack of an inclusive code or standard to fully address the system, indicates that a failure modes effects and criticality analysis (FMECA) was warranted. FMEA is a bottom-up, inductive analytical method that may be performed at either the functional or piece-part level. FMECA extends FMEA by including a criticality analysis, which charts the probability of failure modes against the severity of their consequences and looks at the systems as a whole. In February 2020, "structural engineers [WSP Global, Inc.] were on site and performed inspection of the towers, cables, and platform primary structural elements. No additional damage was noted to have been found during those inspections."[98] Not taking measures such as the following—(1) performing thorough structural assessments by licensed engineers to continue operations after Hurricane Maria in 2017, the earthquake in January 2020, or a specified duration in service (say 50 years); (2) developing and implementing a plan to replace cables after a specified period in service; or (3) developing and implementing a plan to increase the rigor and frequency of inspections with time in service—all may have increased the probability of failure.

Conclusion: The consequences of a structural failure of the Arecibo Telescope were not seriously considered in decision-making during design and operation or in extending the telescope's life. In particular, there was no formal consideration that the health and safety of the workers and the public were at risk in the event of a structural failure. The design to convert it into a telescope with public visitors and the re-design to add the Gregorian dome were not conducted using more stringent standards for critical structures like bridges, even though workers and the public in the Visitor's Center and the Learning Center could have been harmed in the event of a single cable failure or a catastrophic collapse. The potential life and safety consequences associated with a single cable failure or a catastrophic system collapse on workers and the public were not considered when damage was detected to cable M8-4 after Hurricane Maria or in the decision-making to continue operations in 2017.[99]

Finding: The performance of the Arecibo Telescope's cable support system, with multiple lines requiring repair and replacement over its 60-year service life, was not unusual or unprecedented.

Conclusion: The risk of a structural collapse could have been reduced if more rigorous inspections and assessments had been conducted to evaluate the integrity of the cables and connections. These inspections should have been comprehensive and done with an understanding of all potential failure mechanisms, including those time-dependent degradation processes that can operate below 50 percent of the cable-breaking strength.

STRUCTURAL ROBUSTNESS

Structural robustness is defined as the inherent health and strength of the structural system to withstand external demands without degradation or loss of functionality. Some of the measures to enhance robustness include the following:

- Increasing the accuracy and reliability of the design loads;
- Establishing a clear, simple, logical design;
- Providing redundancy in the design;
- Employing experienced and qualified design team, contractors, and inspectors;
- Providing quality assurance over the design, construction, maintenance, and inspection over the life of the facility; and
- Increasing the factor of safety in critical elements.

---

[98] NSF presentation, slide 21.
[99] NSF, 2017, "Environmental Impact Statement for Arecibo Observatory, Arecibo, Puerto Rico," July 27, https://www.nsf.gov/mps/ast/env_impact_reviews/arecibo/eis/FEIS.pdf.

Increasing the factor of safety in the design of the members—or oversizing them—has advantages and disadvantages. Where structural members are over-sized, the over-sized connections themselves may create problems in fabrication and quality assurance. Huge weldments and huge bolted connections are more difficult to build, and maintaining quality may be more difficult. Therefore, the decision to increase the factor of safety should be made carefully, taking into account all the issues that may be affected. For the reasons cited elsewhere in this report, the committee does not agree with TT's recommendation to increase the factor of safety for the Arecibo Telescope. Oversizing in response to technical unknowns is not prudent and may not help if material failure can occur at less than 50 percent of breaking load or yield strength through unappreciated mechanisms such as LEP, hydrogen embrittlement (HE), SCC, or creep. Oversizing can only help by lowering the applied stress below the threshold for the material degradation process.

Redundancy

In structural engineering, a redundant structure is one where the structural system has alternate load paths so that the removal/failure of a member does not initiate a total collapse of the system. Critical members of the structural system should be designed such that one of these members could be removed without the failure of the whole structure. It may be impractical or impossible to design for the simultaneous removal of several or all of the critical members. Considering various scenarios of individual member failures, the structural engineer should propose reasonable approaches, discuss them with the owner and/or stakeholders, and jointly decide on a system.

Based on the committee's experience, a typical approach is to consider the joint probability of several critical members failing simultaneously and then apply a smaller safety factor for these simultaneous failure load cases. Employing such a method ensures that the increase in material quantities is nominal.

More redundancy is required for more critical structures, structures with longer design lives, structures exposed to the environment, structures subject to fatigue, and structures where individual members, due to corrosion or other degradation, must be replaced over time. The designer should consider possible risks and provide sufficient internal redundancy where justified. A structural system with three supporting legs is inherently non-redundant, as removing one of the legs will cause a total collapse.

With a three-legged system, there is no practical way to replace a set of cables (to one tower) that may have been entirely severed for any reason. Applying the redundancy concept, a four-legged system can be designed to provide a higher confidence of tower reliability if cables are severed. Theoretically, in that failure scenario for a four-legged design, the designer could consider the tower loss and calculate additional forces in the remaining cables and check the remaining tower resistance to the increased lateral loads, and design accordingly. Furthermore, the four-legged system allows the installation of a temporary overhead cable system in case of emergencies. Regardless of tower configuration, design for cable replacement is a normal feature of cable-supported structures, and it will require special strength margins or member redundancy.

> The first socket failure occurred at the end of an isolated cable (M4N), whose tension could not be redistributed to adjacent cables. Instead, this first failure resulted in a rotation of the platform and tension changes throughout the cable system. Designing cable systems with multiple adjacent cables on each span provides redundancy. In the event of a cable failure, the remaining adjacent cables can sustain the increased load for some time assuming they have not lost capacity or seen it reduced by degradation, allowing and easing the replacement of the failed cable or cables.[100]

The original design of the three towers had significant reserve capacity, corroborated by the fact that the 1997 upgrade to the telescope added 40 percent to the weight of the suspended structure. Still, it did not require the three towers to be structurally reinforced. The towers seemed to have factors of safety that were higher than required.

The original 1963 design of the telescope did have redundant cable design in each tower (but not the whole system). However, the 1997 upgrade lacked redundancy in its design, as single auxiliary cables were used on either side of each tower to accommodate the additional platform load. Loss of an auxiliary cable would, at a minimum,

---

[100] TT Final Report, p. 50.

substantially rotate and shift the platform, rendering it useless. These single auxiliary cables are M4N and M4S, connecting the platform to Tower 4, and at B4N and B4S, connecting Tower 4 to Anchorage 4.[101] The same design was used at Tower 8 and Tower 12. With this change to the original 1963 design, the whole cable system was no longer redundant, and the reliability of the original cable design was compromised.

The 1997 upgrade should have maintained the cable redundancy of the original design, but it did not. Considering the magnitude of the renovation, it should have brought the entire structural system to conform to the current codes, technologies, and design methodologies of 1997. The lesson for NSF is that facility upgrades should not be allowed to compromise the reliability and redundancy of the existing facility. Fundamentally, it is better design practice for a structure to have alternate load paths (i.e., to have redundancy) than to have over-sized structural members (i.e., members with higher than required factors of safety). Given that the loss of the single M4N cable did not immediately fail the system, it is likely that an operative operations and maintenance program that included cable replacement at given performance thresholds or design life could have precluded failure.

Conclusion: The Arecibo Telescope structure was not tolerant of cable failure due to the lack of redundancy in the 1997 upgrade, where a single auxiliary cable was used on each load path instead of multiple auxiliary cables.

MONITORING

A highly stressed critical structural member should have triggered a higher-than-normal observation program, including some form of definitive, repeatable measurement of socket performance with measurable performance limits prescribed by the designer to facilitate the interpretation of the measurement. Greater urgency for inspection and maintenance, when cable slippage is observed, is warranted. Especially if it is recognized that there are material failure modes that can occur at a fraction of the wire yield strength, such as LEP, SCC, HE, corrosion fatigue, and PLC, which are time-dependent and reduce load capacity over time.

Monitoring Program Used for the Cable Sockets

Throughout the life of the Arecibo Telescope, the condition of the cable system and the maintenance operations were routinely inspected, and cable conditions were recorded locally by AO staff. Occasional structural inspections were also reported by the Engineer of Record, A&W, between 1972 and 2011. The available information is generally less comprehensive and detailed after the second upgrade in 1997, and the scope of the inspections performed by A&W was reduced.[102]

The monitoring program for the sockets used over the life of the tower consisted of periodic visual inspections. There was no systematic methodology for observing and recording the performance of the sockets. A comprehensive review of the inspection history for the structure by TT[103] showed no systematic inspection of the sockets nor any formal monitoring program in which systematic measurements or records were made of the socket performance.

Photos of cable pullout were taken at various times. Examples include Figures 14, 15, 16, 18, 23, 24, 30, 34, and 43 in the TT Final Report. Some of these photos include a ruler showing the amount of pullout at the time of the photo. These illustrate that meaningful measurements of pullout could have been made and tracked systematically. TT reported no evidence of any systematic monitoring or tracking of socket conditions. No records of measurements of slip over time for specific sockets have been found.

The cable slips were measured at the tower and ground ends of the auxiliary cables after the first cable failure and at the platform ends after the collapse, as shown in Figure 3-14. One-third of the cable slips exceeded the AASHTO limit (only after the connection is loaded to 80 percent of its nominal strength) of one-sixth of the cable

---

[101] TT Final Report, Figure 22.
[102] TT Final Report, p. 8.
[103] TT Final Report, Appendix D, p. 12.

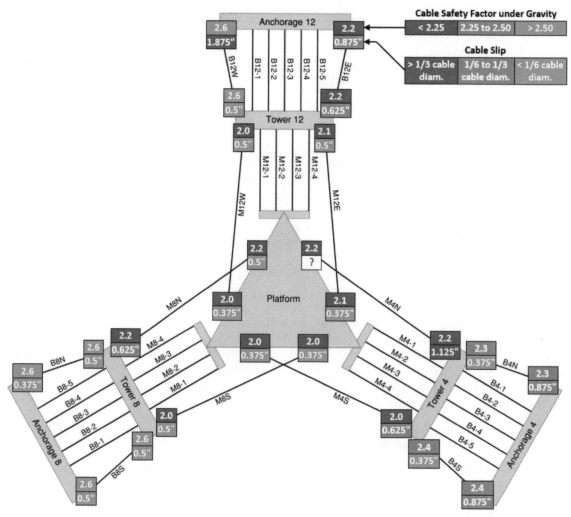

FIGURE 3-14 Cable safety factors and measured cable slips on auxiliary sockets.
SOURCE: Thornton Tomasetti, 2022, *Arecibo Telescope Collapse: Forensic Investigation*, NN20209, prepared by J. Abruzzo, L. Cao, and P.E. Pierre Ghisbain, July 25, https://www.thorntontomasetti.com/sites/default/files/2022-08/TT-Arecibo-Forensic-Investigation-Report.pdf; courtesy of Thornton Tomasetti.

diameter.[104] The maximum cable slip was observed at the ground end of B12W. It had increased from 1.5 inches in May 2018 to more than 1.75 inches in September 2020, which prompted the plans for interim repair with a friction clamp after the first cable failure.[105]

Before breaking free on August 8, 2020, the end of cable M4N at the top of Tower 4 had slipped by more than 1 inch from its socket. This 1-inch slip is approximately one-third of the cable's diameter and significantly more than typically observed in structural cables terminated with zinc-filled spelter sockets. Notably, while this slip was observed and recorded, it did not trigger any extra attention or action.

TT characterized these measurements into three categories: those less than D/6, those between D/6 and D/3, and those greater than D/3. TT does discuss the reasoning for using these limit values, apart from the misapplied D/6 discussed previously. Still, they indicate what should have been done to establish performance criteria for a

---

[104] AASHTO, 2019, "Standard Specification for Wire Rope and Sockets for Movable Bridges," M 277-06.
[105] TT Final Report, Figure 24.

socket health monitoring program. In simple monitoring terms, one often establishes a (ideally nonjudgmental) traffic light system: green for *OK*, yellow for *Caution*, and red for *Take Action*. Using this for illustration, Figure 25 in the TT Final Report would show 2 of 23 sockets in a red state and 6 of 23 in a yellow state. Taken together, 8 of 23 sockets, or 1/3 of the sockets, had displacements after the collapse that could have been considered excessive and warranting detailed examination. (One socket could not be characterized and is not included in this count.)

Role of Performance Monitoring for Critical Structures

In the context of this document, a critical structure is one that poses a high threat to life should it collapse. The Arecibo Telescope met this definition of a critical structure. Performance standards are higher for a critical structure due to their higher risks.

While not regulatorily mandated, good practices in civil engineering usually promote adopting a systematic performance monitoring program to (1) verify that the facility is performing to the design specifications and (2) detect any indications of deteriorating performance that would necessitate actions to mitigate a potential collapse. The need for these practices is especially true for critical structures. Systematic performance monitoring entails conducting and documenting measurements and observations in a predetermined manner at fixed time intervals that enable competent experts to assess the variations in performance over time and to identify any unacceptable levels of risks that require corrective measures. The level and sophistication of a performance monitoring program for a constructed facility usually increase with increasing risks and should increase in frequency as the structure ages.

What Monitoring Was Possible and Practical?

It is not unusual for a cable to slightly displace out of a spelter socket when it is first loaded as the zinc casting seats within the socket's cone, and ⅜ inch was measured in the Arecibo Telescope's socket testing at Lehigh University. It is expected that some displacement out of a spelter socket will occur when the load is applied during construction. Then any further displacement will diminish to a small amount or zero unless a new load is added, such as by wind or earthquake. Monitoring of the extraction pullout of each socket could have been performed and used as a quantitative metric of socket performance. Figure 23 from the TT Final Report illustrates how visible the socket pullout was at the end of each socket.

At least one definitive criterion for the allowable slip of cable at spelter sockets existed based on judgment and experience.[106] This standard applies to moveable bridge cables. The similarity of the two cable systems was enough that other consultants looked to this standard for establishing a maximum pullout limit for the Arecibo Telescope's cables. This standard limits allowable slip to one-sixth of the cable diameter when proof-loaded to 80 percent of the cable's minimum breaking strength, a load level that was never seen by the Arecibo Telescope's cables. Arecibo cable diameter was 3 to 3⅜ inches, so a possible pullout threshold limit was approximately ½ inch. For the telescope's main and backstay cables, this limit corresponds to a maximum slip between 0.5 and 0.6 inches.[107]

A slippage of ½ inch could have been easily recorded by manual means periodically using a camera, aided with a ruler or micrometer for scale, as is shown in some of the figures in the TT Final Report.[108] The ease of recording such slips is demonstrated by Figures 23 and 24 from the TT Final Report, shown below in Figure 3-15. TT Figure 23 shows a 1.125-inch pullout of M4N-T on February 19, 2019 (538 days before it failed). Figure 24 shows an increased pullout of B12W-G of more than ¼ inch between May 15, 2018, and September 18, 2020. Automated monitoring using displacement transducers and data loggers could have been added at any point to obtain more precise data to reveal trends with time and events, possibly gaining insight into the slip mechanism and its correlation with site events. Such measurements with appropriate interpretation could also have provided

---

[106] AASHTO, 2019, "Standard Specification for Wire Rope and Sockets for Movable Bridges," M 277-06.

[107] From TT Final Report, the cables reported max load was approximately 62 percent of their minimum breaking strength so these displacement limits should be reduced to approximately 0.4 to 0.5 inches. A ½ inch amount would have been a reasonable practical limit to set to trigger further attention to the socket.

[108] TT Final Report.

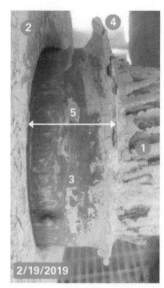 

FIGURE 3-15 Images showing physical measurement of cable socket slippage.
SOURCE: Thornton Tomasetti, 2022, *Arecibo Telescope Collapse: Forensic Investigation*, NN20209, prepared by J. Abruzzo, L. Cao, and P.E. Pierre Ghisbain, July 25, https://www.thorntontomasetti.com/sites/default/files/2022-08/TT-Arecibo-Forensic-Investigation-Report.pdf; modified from photos from NAIC Arecibo Observatory, a facility of the National Science Foundation; courtesy of Thornton Tomasetti.

indicators that contingency measures should be engaged to avoid the collapse or at least steps to be taken to reduce the risks to people and property.

TT concluded,

> It is now clear from our study that excessive cable slip occurs in zinc-filled spelter sockets due to zinc flow and is a sign of upcoming failure through core rupture or core flow-out. Excessive cable slip was observed on the first socket that failed at least a 1½ years before the collapse but was not identified as an immediate structural concern. Monitoring the cable slip and slip rate is a reasonable method to determine if a socket is failing, and the limit of one-sixth of the cable diameter appears to be a reasonable threshold for slip monitoring based on what was observed on the telescope's sockets. Cable slip can cause the rupture of individual outer wires before complete socket failure. A socket exhibiting outer wire ruptures should, therefore, be closely inspected and monitored for cable slip in its connections. However, monitoring wire ruptures is not sufficient, or even directly correlated, to determine if a socket is failing.[109]

Further from TT,

> A safe cable system can still be designed with a lower safety factor, such as the 2.2 safety factor prescribed in ASCE 19. However, in that case, the sockets should be inspected regularly to measure cable slip. A socket should be replaced or bypassed when excessive cable slip indicates that zinc flow continues to occur over time. Limiting the allowable cable slip to one-sixth of the cable diameter is reasonable until further studies are performed.[110]

The following captures key points relative to the monitoring of the cable sockets:

- Cable sockets may be the structure's weak point.
- Cable socket pullout was expected to remain negligible after initial loading.

---

[109] TT Final Report, p. 49.
[110] TT Final Report, p. 50.

- There were multiple indications of excessive cable slip in the sockets.
- The observatory would have benefitted from a systematic monitoring program to monitor cable pullout of the sockets and a contingency plan to deal with unacceptable pullout (a pullout of more than D/6). Pullout of 1.5 inches, increasing to 1.75 inches, was observed in September 2020 on B12W, with more than 1 inch observed on M4N. These exceed D/6 by far and should have immediately triggered a systematic monitoring program post-Maria. Monitoring is meant to include systematic quantified measurements of performance that can be duplicated by an independent person. While visual inspections are an important aspect of monitoring, they are not sufficient to provide definitive quantitative data over time.
- There was never a systematic monitoring program in place for the facility.

Finding: Meaningful measurements, documented with photos, of socket slip could have been obtained with simple equipment such as a ruler or micrometer.

Conclusion: Such measurements at sufficient intervals would have indicated excessive socket slip and that the cable performance was not uniform across the structure.

Conclusion: Differentials in cable slip and/or measurements warranted serious investigation by experts knowledgeable in socket performance. Any recommendations from such experts would have to be implemented.

These recommendations related to critical structure performance monitoring are based on lessons learned from the Arecibo Telescope collapse and the experience of the committee members.

Recommendation: The facility owner/operator should ensure that an operations and maintenance manual for the structure is commissioned and is available during the operation of the structure. The manual should:
- **Identify performance standards of the facility to help detect unexpected, potentially dangerous performance and deteriorating performance with time;**
- **Provide a monitoring and inspection plan that considers potential critical failure modes (and necessary inspection expertise to address them) and include physical variables to monitor, locations to monitor, and the recommended frequency of monitoring. The plan should recognize that some time-dependent failure modes can operate at low loads in contradiction with the safety factor. It should also provide limit values for warning levels and action levels for each performance variable to be monitored. (Warning level is the point where performance becomes concerning, and further evaluation of the safety of the structure should be made. The limit level endurance limit is the point where the performance is becoming threatening to life, and people should be removed from harm's way.); and**
- **Indicate the expected service life of the facility and its key components.**

Recommendation: The facility owner/operator should:
- **Implement the monitoring plan and keep it operational for the life of the structure. For structures with long life expectancies, this may require updating to account for mechanisms and degradation that are a function of age; and**
- **Engage a qualified professional to evaluate the monitoring data at least annually, assess the safety of the structure, and provide recommendations for changes to the structure and changes to the monitoring plan as needed.**

4

Arecibo Telescope's Management and Oversight

Hurricane Maria struck the Arecibo Telescope during the period in which the National Science Foundation (NSF) would have been considering proposals in response to its solicitation for the Management and Operations of the Arecibo Telescope, due on May 4, 2017 (NSF-538), and the award to the University of Central Florida (UCF). Following the award of the Arecibo Telescope operating contract to UCF, the full management contract started on April 1, 2018.[1] The transition period of Arecibo Observatory (AO) to UCF lasted from February 2018 through June 2018, supported by an award for $913,935.[2] SRI International was the managing organization until the end of March 2018, and the organization received an extension to support close out until May 31, 2018, more than 6 months after Hurricane Maria. There were several meetings among NSF, SRI, and UCF during the facility's turnover. The AO site director, as well as the majority of the AO staff, remained throughout this transition. The $20 million grant to fund UCF's operations and maintenance (O&M) award was not received until April 2018,[3] 7 months after Maria had struck the Arecibo Telescope. The UCF proposal separated the Arecibo Telescope's science and maintenance functions, with the responsibility "for facilities infrastructure, engineering, operations, maintenance, information technology and support services, as well as logistics and security services" to be assumed by Yang Enterprises, Inc.[4] While the announcement stated that "Yang Enterprises, Inc. will introduce new technologies and cutting-edge tools to support Arecibo's requirements and modernize operations,"[5] the committee did not find evidence of significant change in the Arecibo Telescope's inspection methods and maintenance practices after UCF assumed control of operations. During the transition period, UCF management planned to implement a computer maintenance management system. Most maintenance records, however, were kept onsite as hard copies.

UCF assumed the Arecibo Telescope's operation in the face of a budget that had been declining over the previous 5 years, and NSF announced a decision in 2017 (a few weeks after Maria) that "operations at the observatory

---

[1] National Science Foundation (NSF), 2023, "Operations and Maintenance of the Arecibo Observatory," NSF Grant Award #1822073, https://www.nsf.gov/awardsearch/showAward?AWD_ID=1822073&HistoricalAwards=false.

[2] NSF, 2018, "Statement on Award of Cooperative Agreement for Management and Operations of Arecibo Observatory," February 22, https://www.nsf.gov/news/news_summ.jsp?cntn_id=244536&org=NSF&from=news.

[3] Z. Kotala, 2019, "UCF Marks 1st Year at Arecibo Observatory in Puerto Rico," *UCF TODAY*, University of Central Florida, April 5, https://www.ucf.edu/news/ucf-marks-1st-year-arecibo-observatory-puerto-rico.

[4] N.T. Tillman, 2018, "Arecibo Observatory Announces Identity of 'Mysterious Benefactors,'" *Space.com*, February 27, https://www.space.com/39827-arecibo-observatory-new-management.html.

[5] Ibid.

[continue] with reduced agency funding."[6] However, when partner funding failed to materialize to make up the difference, NSF maintained the O&M budget. Eight months after Hurricane Maria, NSF awarded UCF $2 million to be focused on time-critical repairs.[7] In 2019, "NSF submitted waiver to OMB to permit $11.3M of $14.3M to be spent over 60 months."[8] Following the hurricane, repairs began in December; however, they could not access the platform because of the catwalk damage.

Both through solicitations and oversight, NSF requires the contractor to take ownership of the maintenance and operation of the facility.[9] However, maintenance, inspection, and repairs are terms often misunderstood and thus not correctly applied when justifying funding for the operation of aging and dated facilities. Based on the committee's professional experience, this lack of appreciation for the importance of maintenance, inspection, and repair is especially true for large-scale assets that have become technologically dated for their original intended purposes and are no longer state of the art. Financial administrators may regard maintenance and inspection as optional expenses that can be deferred while regarding repairs as expenses, and even then, only as long as the damaged items needing repair negatively affect the main operation of the facility and thus cannot be postponed. Decisions to reduce operations and maintenance funding in favor of other activities assume some risk to the health and performance of the structure. The challenge, then, is that during times of managerial change, the new personnel may not be fully informed of these assumed risks. Therefore, if a low-probability event does occur, it may happen to staff unaware of the risk.

Recommendation: The National Science Foundation and organizations that use similar site management contracts to manage their portfolios should consider funding for the inspection, monitoring, maintenance, and repair of aging facilities and infrastructure as important as they are critical to the structure's performance and longevity.

For any complex, complicated, and unique facility operated by NSF, such as the Arecibo Telescope, inspection, monitoring, maintenance, and repairs must occur at two levels. First and most obvious, a Level One maintenance, inspection, and repair program should apply to the operational components of the facility, such as the antennas, receivers, and mechanical and electronic equipment. The malfunctioning of any one of these components is readily noticed and addressed by the operators of the facility in charge of conducting the research activities. The facility's equipment items are often supplied by a manufacturer who typically provides maintenance, inspection, and repair manuals for their equipment. Furthermore, these manuals also provide guidance for addressing probable malfunctioning and troubleshooting problems and verifying correct operation. These manufacturers have the most knowledge of all the design assumptions, risks, and unknowns regarding the performance of their equipment. Other aspects of this Level One maintenance, inspection, and repair of a facility include typical tasks such as painting, corrosion staining, disconnected cables, etc., which do not require an understanding of the complexity of the facility, just experience with typical daily maintenance of any facility.

Level Two maintenance, inspection, and repair of the facility is often problematic unless there is a specific and informed delegation of this responsibility. NSF facilities with long life expectancies, like the Arecibo Telescope, often incorporate or upgrade with state-of-the-art technologies and materials that, at the time of design and construction, were justified as a calculated risk to obtain other benefits. By the very nature of advanced design, long-term performance has not been demonstrated under the design conditions. Experience is being gained through the application of many continuously evolving technologies, and thus, their performance must be verified with time.

---

[6] NSF, 2017, "Arecibo: Statement on NSF Record of Decision," November 16, https://www.nsf.gov/news/news_summ.jsp?cntn_id=243729.

[7] NSF, 2018, Award 1838728, "Hurricane Maria Relief Funding for Restoration of Scientific Capabilities at the Arecibo Observatory," May, https://www.nsf.gov/awardsearch/showAward?AWD_ID=1838728.

[8] A. VanderLey, 2022, "Arecibo Observatory: Failure Event Sequence," National Science Foundation presentation to the committee, January 25, slide 13.

[9] See the following NSF Program Solicitations: NSF 10-562, "Management and Operations of the National Astronomy and Ionosphere Center (NAIC)," posted April 29, 2010, https://new.nsf.gov/funding/opportunities/nrao-management-operation-national-radio-astronomy-observatory/503352/nsf10-562/solicitation, and NSF 17-538, "Management and Operations of the Arecibo Observatory," posted January 25, 2017, https://new.nsf.gov/funding/opportunities/management-operations-arecibo-observatory/505401/nsf17-538/solicitation.

The Design Engineer or Engineer of Record (EOR) for the original design or significant modifications/upgrades is most knowledgeable concerning the risks, technologies, assumptions, redundancies, weaknesses considered in the design, and the potential threats to overall structural integrity. The EOR generally has knowledge of proof testing, field validation, and quality control in design and construction, what critical structural components need to be monitored, locations to inspect, and the limits of adequate performance of the critical components within the context of their design scope. Unfortunately, the original EOR is often not involved in significant changes or upgrades, particularly when there is a significant time lag involved. In these cases, the EOR for the changes must assume this role for all parts of the structure that are impacted by modifications, even if they were not modified.

The most recent EOR should provide a written narrative for the facility records describing the final conceptual design and its expected performance, as well as identify the critical components affecting the structure's performance and how these should be monitored. This step is doubly important for a facility whose planned operation is by competitively bidding contractors who will have had no role in the facility design. Furthermore, the EOR should be retained to develop an inspection and monitoring manual for the ultimate facility operator, clearly outlining the procedures to monitor the performance of the most critical components affecting the facility's structural integrity. To the extent possible, the prescribed monitoring should not require engineering knowledge or judgment from the inspectors and instead prescribe physical monitoring of non-judgmental parameters, such as counted wire breaks or measured cable pullouts. The inspection manual should also clearly identify the acceptable limits for the facility's operation and inspection actions to be taken when the operation, or degradation such as a cable pullout, exceeds the acceptable limits, including safety measures to ensure the safety of the personnel, employees, scientists using the facility, and the visiting public. Online structural monitoring and health management, condition monitoring, condition-based maintenance, and reliability-centered maintenance are typical of today's monitoring techniques.

The inspection, monitoring, maintenance, and repair program of the structural portion of the facility developed by the EOR should include clear guidelines for reporting and evaluating the information in the reports and records from both the Level One and the Level Two programs. Of utmost importance is the fact that the inspection, monitoring, maintenance, and repair program should include guidelines for reporting and assessing the criticality of any structural findings, especially those considered unusual or unexpected. The Arecibo Telescope lacked the proper inspection, monitoring, maintenance, and repair manuals addressing critical structural considerations that only the EORs in the original design and the upgrade knew. This deficiency is most significant when considering that the expected service life for the facility was 10 years.[10] The Arecibo Telescope underwent several changes, which significantly increased the applied loads. At the time of the collapse, it had been in service for 57 years. NASA concluded, "The design did not explicitly consider the time-dependent effects of creep and cyclic loading on design capability, nor did it set service life inspection intervals with pass/fail inspection criteria. It also did not specify an end-of-life capability requirement associated with service life degradation."[11] All of these deficiencies resulted in a facility that had exhibited unheeded signs of distress for several years.

The collapse on December 1, 2020, resulted from the lack of recognition of the significance of the cable pullout by the experts contracted to assess the Arecibo Telescope and to develop a response plan after Hurricane Maria. There was inadequate funding for the repairs or replacements needed because of the cable pullout. The operating entity and its managers failed to consider all the findings from the facility inspections to ensure the safety of its employees, scientists, staff, and the public visiting the facility.

The following findings and recommendations are relevant to any large structural facility owned by NSF with contract operators.

> Finding: A Level Two inspection, maintenance, monitoring, and repair manual was not developed by the EOR or NSF, nor was such a policy implemented during the management of the Arecibo Telescope by its last two operators.

---

[10] Interview of William Gordon by Andrew Butrica, Niels Bohr Library & Archives, American Institute of Physics, College Park, MD, November 28, 1994, http://www.aip.org/history-programs/niels-bohr-library/oral-histories/22789.

[11] NESC Report, p. 120.

Conclusion: A good practice would be to have separate accounting and funding, independent of facility operations, for inspection, maintenance, monitoring, and repair of facilities. Otherwise, inspection, maintenance, monitoring, and repair may suffer in financial competition with the science operations that can gain the facility further funding.

Recommendation: Facility owners should enforce compliance with contract requirements through independent auditing of inspection, monitoring, maintenance, and repair records.

The Arecibo Telescope's condition monitoring, while spotty throughout its history, demonstrably degraded after the 1997 upgrade and further degraded after 2011. The first comprehensive Ammann & Whitney inspection of the telescope after the completion of the Gregorian upgrade in 1997 was in 2003. The following comprehensive inspection was not until late 2010. While there were continuous inspections, apart from broken wire counts, these were generally only visual inspections, and photographic documentation is irregular. In the obsolete system, there was little codification or organization of the data with which to observe any trends. There appear to have been no specified procedures as to who should be contacted if there were any serious or discrepant findings. After supervisors were informed, there was no process to involve the chief engineer for additional analysis. There was also no central repository of the maintenance and inspection data at the facility, as shown when UCF began management of the facility, and UCF contemplated visiting Cornell University to review the archives.[12] Even though the Arecibo Telescope was aging, inspection methodologies did not change and seem to have become less frequent, as indicated by the reduced available inspection data discussed previously. By the time extensometers were considered, it was too late. Cable and socket maintenance consisted of applying sacrificial zinc paint; epoxy paint was used everywhere else.

---

[12] R. Lugo and F. Cordova, 2022, "Perspectives on Grant Award and Operations of Arecibo Observatory Cooperative Agreement by the University of Central Florida," University of Central Florida presentation to the committee, February 17, meeting transcript minute 53:13.

5

Other Lessons Learned

STATE OF KNOWLEDGE

Transfer of Institutional Knowledge

The Arecibo Telescope is owned by the National Science Foundation (NSF) and was managed by Cornell University from its opening in 1963 to 2011. Management was transferred to a consortium led by SRI International in 2011 and to another consortium led by the University of Central Florida (UCF) in 2018.[1]

Cornell University had to make changes in leadership, staff, and administration working on the Arecibo Telescope for nearly 50 years. How much knowledge and experience with the design and performance of the telescope structure was transferred both during internal organizational changes and later to SRI International and subsequently UCF is unclear. Certain performance aspects, such as temperature, oscillations, wire breaks, and tiedown stresses of the telescope, were monitored. Ultimately, these metrics alone do not paint a complete picture of the cause of the failure.

Hurricane Irma occurred on September 7, 2017, and Hurricane Maria was on September 20, 2017. The transition to the new management under UCF in the first half of 2018 happened at a critical time in the structure's history, both in terms of funding and the impact of the hurricanes on the structure.

It seems unlikely that UCF had adequate time and resources to review and understand the Arecibo Telescope's original 1963 design, the 1974 upgrade, the 1997 upgrade, the structural inspection and maintenance records produced for nearly 50 years, the performance over time, the critical aspects, and the key factors, such as the wire breaks and cable pullout of the sockets and their significance on the strength and integrity of the structure.

In a 2003 inspection, the telescope's Engineer of Record observed that cables had slipped out of their sockets by as much as one-half of an inch and attributed the slips to have occurred during the fabrication or testing of the cables. A later report in 2011 noted that the cable slips remained unchanged from 2003. To our knowledge, the Observatory staff was not instructed to monitor cable slips, nor were they provided a limit on acceptable cable slip. After Hurricane Maria in 2017, the Observatory staff observed and recorded cable slips of more than one inch on two of the sockets.

---

[1] Thornton Tomasetti, Inc., 2022, *Arecibo Telescope Collapse: Forensic Investigation*, NN20209, prepared by J. Abruzzo, L. Cao, and P.E. Pierre Ghisbain, July 25, https://www.thorntontomasetti.com/sites/default/files/2022-08/TT-Arecibo-Forensic-Investigation-Report.pdf (hereafter "TT Final Report"), p. 2.

There is no documentation to show whether these cable slips increased during the hurricane, and to our knowledge, they were not identified as an immediate structural integrity issue.[2]

From this record, one may conclude that the telescope's Engineer of Record, the Observatory staff, and the staff of UCF did not recognize the significance of the cable pullout. The measured cable pullout may have appeared "normal" to them and was not on their radar as signs of structural distress. The lack of concern may be because a small cable pullout was present from the beginning, and no one in authority had previously raised an alarm:

- February 19, 2019: A pullout of 1.125 inches of the cable was measured at the location of M4N-T. However, it would seem that nothing was done about this excessive pullout.
- August 10, 2020: First socket failure. About 18 months after the excessive pullout of the cable was measured, the cable pulled entirely out of the socket at M4N-T.

Finding: Record keeping related to telescope organizational and structural performance was often inconsistent and incomplete over time, including through operational contract transfers.

Documentation

A facility's "owner's manual" should be updated and passed on to other generations of owners/stakeholders and engineers-of-record of the facility. Innovative, unusual, one-of-a-kind structures, long-span, and high-rise structures, particularly, require such documentation. The documentation should provide a list of the governing design building codes and regulations, a detailed description of the structural system, the design life span, identify the critical elements of the structural system, list the strength and types of materials used, the design forces from gravity (dead and live), lateral loads from wind, earthquakes, soil, and water, the design wind speeds for strength and serviceability, the load combinations, the design programs used, modeling assumptions, the building periods, and results of various code checks. The redundancy of the system should be documented, i.e., the various alternate load paths for various scenarios and the reduced factors of safety used in conjunction with these "what if failure" mode cases.

Additionally, a list of signs of distress to be observed at regular intervals by inspectors should be provided. For example, this list might include buckling of structural members, excessive settlements of foundations, leaning of the structure, excessive stretching of tensile elements, cracking in the structure (beams, columns, slabs, trusses, etc.), cable slips, and wire breaks.

Beyond the need for regular inspections, there needs to be timely maintenance and repairs as required based on the inspection reports.

CONTINUED RESEARCH

Forensic testing was performed on only a fraction of the cables and sockets recovered at Arecibo. Additional sockets and wire sections were retained by Thornton Tomasetti, Inc., and are temporarily being stored. Additional connections with measured slippage are available as well as untested platform connections. Research could be conducted on those sockets that could be informative to design and materials science. These sections may have tremendous value to the research community in evaluating the low-current electroplasticity hypothesis proposed in the report; materials testing of multi-decade, in-service, large-diameter steel cable and zinc; or further analysis of brooming.

Recommendation: While still available, the National Science Foundation should offer the remaining socket and cable sections to the research community for continued fundamental research on large-diameter wire connections, the long-term creep behavior of zinc spelter connections, and materials science.

---

[2] TT Final Report, p. ii.

Bibliography

AASHTO (American Association of State Highway and Transportation Officials). 2019. "Standard Specification for Wire Rope and Sockets for Movable Bridges." M 277-06.

Abbaschian, R., L. Abbaschian, and R. Reed-Hill. 2009. *Physical Metallurgy Principles*. Fourth Edition. CENGAGE Learning.

Abruzzo, J., L. Cao, and P. Ghisbain. 2021. "Arecibo Observatory: Stabilization Efforts and Forensic Investigation." Presentation to the Committee on Analysis of Causes of Failure and Collapse of the 305-Meter Telescope at the Arecibo Observatory. February 17.

A&W (Ammann & Whitney). 2011. *Arecibo Radio Telescope Structural Condition Survey*. 2011. Cornell University archives, Arecibo Ionospheric Observatory records, 1958–2010. Arecibo Ionospheric Observatory records, #53-7-3581. Division of Rare and Manuscript Collections, Cornell University Library. Box 37. Folder 8. March.

ANSI (American National Standards Institute). 2020. "ASTM G101: Standard Guide for Estimating the Atmospheric Corrosion Resistance of Low-Alloy Steels."

ANSI/ASCE (American Society of Civil Engineers). 1996. "Minimum Design Loads for Buildings and Other Structures." ANSI/ASCE 7-95. June 6. American Society of Civil Engineers. https://doi.org/10.1061/9780784400920.

ANSI/SEI (Structural Engineering Institute). 2022. "Minimum Design Loads and Associated Criteria for Buildings and Other Structures." ASCE/SEI 7-22. November 30, 2021. American Society of Civil Engineers. https://doi.org/10.1061/9780784415788.fm.

ASCE. 2016. "Structural Applications of Steel Cables for Buildings." ASCE/SEI 19-16.

ASCE. 2023. "ASCE 7 Hazards Report: Standards ASCE/SEI 7-22 – Risk Category III." https://asce7hazardtool.online.

ASCE. 2023. "ASCE 7 Hazards Report: Standard ASCE/SEI 7-22 – Risk Category IV." https://asce7hazardtool.online.

ASTM (American Society for Testing and Materials). 1993. "Annual Book of ASTM Standards: Wear and Erosion; Metal Corrosion." Vol. 03.02. ASTM International.

ASTM. 1998. "ASTM A586-98: Standard Specification for Zinc-Coated Parallel and Helical Steel Wire Structural Strand and Zinc-Coated Wire for Spun-In-Place Structural Strand." 10.1520/A0586-98.

Bradon, J., R.C. Chaplin and I. Ridge. 2001. "Analysis of Resin Socket Termination for a Wire Rope." *Journal of Strain Analysis for Engineering Design* 36(1):71–88. https://doi.org/10.1243/0309324011512621.

Bumgardner C.H., B.P. Croom, N. Song, Y. Zhang, and X. Li. 2020. "Low Energy Electroplasticity in Aluminum Alloys." *Materials Science and Engineering: A* 798. https://doi.org/10.1016/j.msea.2020.140235.

Conrad, H. 1998. "Some Effects of an Electric Field on the Plastic Deformation of Metals and Ceramics." *Materials Research Innovations* 2(1):1–8. https://doi.org/10.1007/s100190050053.

Cooke, W. 1976. "Arecibo Radio Antenna." *IEEE Antennas and Propagation Society Newsletter* 18(5):6–8. https://doi.org/10.1109/MAP.1976.27265.

Courtney, T.H. 1990. *Mechanical Behavior of Materials*. McGraw-Hill.

Cucasi, J.D., P.R. Sere, C.I. Elsner, and A.D. Sarli. 1999. "Control of the Growth of Zinc–Iron Phases in the Hot-Dip Galvanizing Process." *Surface and Coatings Technology* 122(1):21–23.

Doundoulakis, H. 1961. "Radio Telescope Having a Scanning Feed Supported by a Cable Suspension Over a Stationary Reflector." U.S. Patent US3273156.

Enos, D.G., and J.R. Scully. 2002. "A Critical-Strain Criterion for Hydrogen Embrittlement of Cold-Drawn, Ultrafine Pearlitic Steel." *Metallurgical and Materials Transactions A* 33:1151–1166. https://doi.org/10.1007/s11661-002-0217-z.

Fontaine, E., A. Kilner, C. Carra, D. Washington, et al. 2014. "Industry Survey of Past Failures, Pre-emptive Replacements and Reported Degradations for Mooring Systems of Floating Production Units." OTC 25273. Houston, Texas.

Frost, H.J., and M.F. Ashby. 1982. *Deformation-Mechanism Maps: The Plasticity and Creep of Metals and Ceramics*. Pergamon Press.

Harrigan, G.J., A. Valinia, N. Trepal, P. Babuska, and V. Goyal. 2021. *Arecibo Observatory Auxiliary M4N Socket Termination Failure Investigation*, NASA/TM–20210017934 and NESC-RP-20-01585. June 15. NASA Engineering and Safety Center, Langley Research Center. https://ntrs.nasa.gov/api/citations/20210017934/downloads/20210017934%20FINAL.pdf.

Hou, W., and C. Lang. 2004. "Atmospheric Corrosion Prediction of Steels." *Corrosion* 60(3):313–322.

Interview of William Gordon by Andrew Butrica. November 28, 1994. Niels Bohr Library & Archives, American Institute of Physics. College Park, MD. www.aip.org/history-programs/niels-bohr-library/oral-histories/22789.

ISO (International Organization for Standardization). 2012. "ISO 9223: Corrosion of Metals and Alloys—Corrosivity of Atmospheres—Classification, Determination and Estimation." https://www.iso.org/obp/ui/#iso:std:iso:9223:ed-2:v1:en.

Jones, D.A. 1995. *Principles and Prevention of Corrosion*. 2nd Edition. Pearson.

Kavanaugh, T.C., and D.H.H. Tung. 1965. "Arecibo Radar-Radio Telescope – Design and Construction." *Journal of the Construction Division* 91(1):69–98.

Komura, T., K. Wada, H. Takano, and Y. Sakamoto. 1990. "Study into Mechanical Properties and Design Method of Large Cable Sockets." Structural Eng./Earthquake Eng. 7(2):251s–262s. October. Society of Civil Engineers (Proc. of JSCE No. 422/1-14).

Kotala, Z. 2019. "UCF Marks 1st Year at Arecibo Observatory in Puerto Rico." *UCF TODAY*. News Release. April 5. University of Central Florida. https://www.ucf.edu/news/ucf-marks-1st-year-arecibo-observatory-puerto-rico.

Lahiri, A., P. Shanthraj, and F. Roters. 2019. "Understanding the Mechanisms of Electroplasticity from a Crystal Plasticity Perspective." *Modelling and Simulation in Materials Science and Engineering* 27(8). https://doi.org/10.1088/1361-651X/ab43fc.

Lugo, R., and F. Cordova. 2022 "Perspectives on Grant Award and Operations of Arecibo Observatory Cooperative Agreement by the University of Central Florida." Presentation to the Committee on Analysis of Causes of Failure and Collapse of the 305-Meter Telescope at the Arecibo Observatory. February 17.

Machlin, E.S. 1959. "Applied Voltage and the Plastic Properties of "Brittle" Rock Salt." *Journal of Applied Physics* 30:1109–1110. https://doi.org/10.1063/1.1776988.

Margot, J.L., A.H. Greenberg, P. Pinchuk, A. Shinde, et al. 2018. "A Search for Technosignatures from 14 Planetary Systems in the Field with the Green Bank Telescope at 1.15-1.73 Ghz." *Astronomical Journal* 155(5). https://doi.org/10.3847/1538-3881/aabb03.

Moen, B.E., O.J. Møllerløkken, N. Bull, G. Oftedal, and K.H. Mild. 2013. "Accidental Exposure to Electromagnetic Fields from the Radar of a Naval Ship: A Descriptive Study." *International Maritime Health* 64(4).

Morales, J.C., and Suárez, L.E. 2020. "Collapse of the Arecibo Observatory in Puerto Rico: Reflections from the Structural Engineering Perspective." *Revista Internacional de Desastres Naturales, Accidentes e Infraestructura Civil* 19–20(1).

NSF (National Science Foundation). 2006. *From the Ground Up: Balancing the NSF Astronomy Program*. D.C. Report of the National Science Foundation Division of Astronomical Sciences Senior Review Committee. Arlington, VA.

NSF. 2017. " Arecibo: Statement on NSF Record of Decision." News Release. November 16. https://www.nsf.gov/news/news_summ.jsp?cntn_id=243729.

NSF. 2018. "Statement on Award of Cooperative Agreement for Management and Operations of Arecibo Observatory." News Release. February 22. https://www.nsf.gov/news/news_summ.jsp?cntn_id=244536&org=NSF&from=news.

NSF. 2023. "Operations and Maintenance of the Arecibo Observatory." NSF Grant Award #1822073. https://www.nsf.gov/awardsearch/showAward?AWD_ID=1822073.

NSF. n.d. "Arecibo: Facts and Figures." https://www.nsf.gov/news/special_reports/arecibo/Arecibo_Fact_Sheet_11_20.pdf. Accessed June 1, 2023.

NSF. n.d. "Report on the Arecibo Observatory, Arecibo Puerto Rico Required by the Explanatory Statement Accompanying H.R. 133, Consolidated Appropriations Act, 2021." https://www.nsf.gov/news/reports/AreciboReportFINAL-Protected_508.pdf. Accessed June 1, 2023.

Phoenix, L., H.H. Johnson, and W. McGuire. 1986. "Condition of Steel Cable After Period of Service." *Journal of Structural Engineering* 112(6):1263–1279.

PTI (Post-Tensioning Institute). 2018. "PTI DC45.1-18: Recommendations for Stay Cable Design, Testing, and Installation." DC-45 Stay Cable Bridge Committee.

Rashad, R. 2005. *Dynamics of the Arecibo Telescope*. Department of Mechanical Engineering. McGill-Queen's University Press.

Ridge, I., and R. Hobbs. 2012. "The Behaviour of Cast Rope Sockets at Elevated Temperatures." *Journal of Structural Fire Engineering* 3(2):155–168. https://doi.org/10.1260/2040-2317.3.2.155.

Siemion, A.P.V., D. Werthimer, D. Anderson, H. Chen, et al. 2011. "Developments in the Radio Search for Extraterrestrial Intelligence." *2011 XXXth URSI General Assembly and Scientific Symposium*. https://doi.org/10.1109/URSIGASS.2011.6051263.

Smedley, P., and D. Petruska. 2014. "Comparison of Design Requirements and Failure Rates for Mooring Systems." Proceedings of the Offshore Structural Reliability Conference. American Petroleum Industry, Houston, TX.

Spence. J.W., F.H. Haynie, F.W. Lipfert, S.D. Cramer, and L.G. McDonald. 1992. "Atmospheric Corrosion Model for Galvanized Steel Structures." *Corrosion* 48(12):1009–1019. https://doi.org/10.5006/1.3315903.

Stavros, A.J. 1987. "Corrosion." Volume 13 in *ASM Metals Handbook*. 9th Edition. ASM International.

Taylor, P.A., and E.G. Rivera-Valentín. 2021. "Fall of an Icon: The Past, Present, and Future of Arecibo Observatory." *Lunar and Planetary Information Bulletin*. Issue 165. Lunar and Planetary Institute. https://www.lpi.usra.edu/publications/newsletters/lpib/new/publication/issue-165-of-the-lunar-and-planetary-information-bulletin-july-2021.

Thompson, T.W., B.A. Campbell, D. Benjamin, and J. Bussey. 2016. "50 Years of Arecibo Lunar Radar Mapping." *URSI Radio Science Bulletin* (357):23–35.

Thornton Tomasetti, Inc. 2021. *Arecibo Telescope Collapse: Forensic Investigation Interim Report*. NN20209. Prepared by J. Abruzzo and L. Cao. San Francisco, CA. November 2.

Thornton Tomasetti, Inc. 2022. *Arecibo Telescope Collapse: Forensic Investigation*. NN20209. Prepared by J. Abruzzo, L. Cao, and P. Ghisbain. San Francisco, CA. July 25. https://www.nsf.gov/news/special_reports/arecibo/Arecibo-Telescope-Collapse-Forensic-Investigation-508c.pdf.

Tillman, N.T. 2018. "Arecibo Observatory Announces Identity of 'Mysterious Benefactors.'" *Space.com*. February 27. https://www.space.com/39827-arecibo-observatory-new-management.html.

Troitskii, O., and V. Likhtman. 1963. "The Effect of Anisotropy and γ Radiation on the Deformation of Zinc Single Crystals in the Brittle State." *Doklady Akademii Nauk SSSR* 148:332–334 (in Russian).

Uri, J. 2022. "65 Years Ago: The International Geophysical Year Begins." NASA History. K. Mars, ed. Updated July 5, 2022. https://www.nasa.gov/feature/65-years-ago-the-international-geophysical-year-begins.

Van Atta, R.H., S.G. Reed, and S.J. Deitchman. 1991. "DARPA Technical Accomplishments Volume II: A Historical Review of Selected DARPA Projects." Prepared for Defense Advanced Research Projects Agency. Institute for Defense Analyses. April. https://apps.dtic.mil/sti/pdfs/ADA241725.pdf.

VanderLey, A. 2022. "Arecibo Observatory: Failure Event Sequence." National Science Foundation presentation to the Committee on Analysis of Causes of Failure and Collapse of the 305-Meter Telescope at the Arecibo Observatory. January 25.

"Wire Cables of Various Types and Materials Tested by U.S. Bureau of Standards," *Engineering Record*, Vol. 72. November 6, 1915. McGraw Publishing.

WJE (Wiss, Janney, Elstner Associates, Inc.). 2021. *Arecibo Observatory: Auxiliary Main Cable Socket Failure Investigation—Final Report*. WJE No. 2020.5191.

Zastrow, M. 2021. "The Rise and Fall of Arecibo Observatory: An Oral History." *Astronomy.com*. September 8. https://www.astronomy.com/science/the-rise-and-fall-of-arecibo-observatory-an-oral-history.

Appendixes

A

Committee Member Biographical Information

ROGER L. McCARTHY (*Chair*) is an independent Mechanical Engineering consultant, an officer and treasurer of the National Academy of Engineering (NAE), and a director on the board of the National Academies Corporation. He also served (2006–2024) as a director on the board of Shui on Land, Ltd., which is publicly traded on the Hong Kong Exchange. Dr. McCarthy was formerly employed by Exponent, Inc. (1978–2008), headquartered in Menlo Park, California, formerly Failure Analysis Associates, Inc. During his more than 30-year tenure, he was variously chief executive officer, chairman, and chairman emeritus. In 1992, then-President Bush appointed Dr. McCarthy to a 2-year term on the President's Commission on the National Medal of Science. He was the commencement speaker for the University of Michigan College of Engineering graduation in April 2008. Dr. McCarthy has investigated some of the major disasters of the current age, most recently in the NAE investigation "*Deepwater Horizon* Explosion, Fire, and Oil Spill in the Gulf of Mexico" for the Secretary of the Interior Salazar. He has appeared on the *History Channel*, *MythBusters*, *Discovery*, *Modern Marvels*, and the *National Geographic Channel*. Dr. McCarthy holds a PhD from the Massachusetts Institute of Technology (MIT) in mechanical engineering.

RAMÓN L. CARRASQUILLO[1] was the founder, president, and consulting engineer working with Carrasquillo Associates (Austin, Texas) and Carrasquillo Engineering Services Group, PSC (San Juan, Puerto Rico). He was formerly a full professor and researcher (1980–2000) at the Department of Civil Engineering at The University of Texas at Austin (UT Austin) and also the founder and president from 1994–2004 of Rainbow Materials, Inc., a ready-mix concrete supply company operating in Central Texas. For more than 20 years, he performed extensive research in the areas of structural engineering, concrete materials, concrete durability, repair and rehabilitation of structures, construction, forensic engineering, failure analysis, and nondestructive testing. Furthermore, since 1980 he provided professional consulting services in the areas of structural engineering, construction, and concrete materials both in the United States and other countries worldwide. He received numerous awards and recognitions, including, among others, the American Concrete Institute (ACI) Foundation Award, the ACI Robert E. Philleo Award, the ACI Henry C. Turner Medal, the UT Austin Faculty Excellence Award, and the T.Y. Lin Award. He was a member of the ACI board of directors, a member and fellow of ACI, and a member of the American Society for Testing and Materials, the National Society of Professional Engineers, the Structural Engineers Society of Texas, the Texas Society of Professional Engineers, and the Puerto Rico Society of Professional Engineers. He was licensed as a Professional Engineer in Texas and Puerto Rico.

---

[1] Deceased on February 2, 2024.

DIANNE CHONG was the vice president of the Boeing Research and Technology organization in the Boeing Engineering, Operations and Technology organization. In this position, she led special projects that impacted processes and program integration for the Boeing Enterprise. Dr. Chong was elected to the NAE and the Washington State Academy of Science in 2017. She currently serves on the National Academies of Sciences, Engineering, and Medicine's Gordon Prize committee; the steering committee for EngineerGirl; the Defense Materials, Manufacturing, and Infrastructure Committee; and the Inclusive, Diverse, and Equitable Engineering for All Committee. Dr. Chong is the immediate past president of ABET. She has served on the board, the COVID task group, and the Inclusion, Diversity, and Equity Committee of the Washington State Academy of Science. She is a member of TMS (The Minerals, Metals, and Materials Society), the American Institute of Aeronautics and Astronautics (AIAA), ASM International, SME, the Society of Women Engineers, Beta Gamma Sigma, and Tau Beta Pi. She has served on the board of trustees and is a fellow of ASM International, and in 2007–2008, was elected as the first female president of ASM International. She has led the ASM Women in Materials Engineering Committee and serves on the ASM Action in Education Team. Dr. Chong is currently the president of SME and is a fellow of SME. She was also inducted into the Women in Manufacturing Hall of Fame in 2020 as part of the inaugural class. She has been recognized for managerial achievements and as a diversity change agent. Dr. Chong has also received an award for alumna achievement from the University of Illinois in 2019. She has also supported national efforts by supporting teams for the National Academies' studies on reproducibility and replicability and materials and manufacturing workshops. Dr. Chong received her bachelor's degrees in biology and psychology from the University of Illinois. She also earned master's degrees in physiology and metallurgical engineering. In 1986, Dr. Chong received her doctorate in metallurgical engineering from the University of Illinois. She also completed an executive master of manufacturing management at Washington University and has a green belt in Six Sigma.

ROBERT B. GILBERT is the chair of the Department of Civil, Architectural and Environmental Engineering at UT Austin. He joined the faculty in 1993 after practicing as a geotechnical engineer for 5 years with Golder Associates, Inc. His technical focus is the assessment, evaluation, and management of risk for civil engineering systems. He has participated in the design or forensic analysis of foundations for a variety of large facilities throughout the world, including offshore energy production systems, bridges, chemical processing plants, and flood protection systems. Dr. Gilbert has been awarded the Norman Medal from the American Society of Civil Engineers (ASCE) and an Outstanding Civilian Service Medal from the United States Army Corps of Engineers. He is a member of the NAE. He served previously on National Academies' committees to assess the performance of waste containment systems and the need for rupture mitigation valves on existing pipelines carrying gas and hazardous liquids.

W. ALLEN MARR, JR., founded and led Geocomp, one of the foremost providers in the United States of real-time, web-based performance monitoring of civil engineering structures, including dams, levees, deep excavations, retaining walls, tunnels, buildings, bridges, and utilities. He now serves as the senior strategic advisor. Dr. Marr also has extensive experience in testing to measure the mechanical properties of earthen materials, designing earth structures, determining the causes of poor performance of geotechnical structures, and developing cost-effective remedial measures for troubled projects. He has been instrumental in applying methods of risk assessment and risk management to infrastructure projects. He has participated in several major failure investigations, including New Orleans levees during Hurricane Katrina, Millennium Tower in San Francisco, and several dams. He is an elected member of the NAE and the Moles. He has published widely and given invited keynote lectures around the world on topics in geotechnical engineering, performance monitoring, data management, and risk management. He served two terms as president of the ASCE Academy of GeoProfessionals. He was awarded the ASCE Seed Medal in 2019 for his contributions to the advancement of research, practice, teaching, and professional service in geotechnical engineering.

JOHN R. SCULLY is the former department chair, the Charles Henderson Named Chaired Professor of Materials Science and Engineering, and the co-director for the Center for Electrochemical Science and Engineering at the University of Virginia (UVA). He received his PhD in materials science and engineering from Johns Hopkins University in 1987. He has had appointments with the Naval Ship Research and Development center, AT&T Bell Laboratories, and Sandia National Laboratories prior to joining the faculty at UVA in 1990. Dr. Scully's technical

APPENDIX A

focus includes most forms of corrosion and environment-assisted cracking in numerous environments, including marine, nuclear, transportation, infrastructure, and aerospace, focusing on a wide variety of materials. He has published more than 340 archival publications on these topics with more than 25,000 citations. He is a recognized scholar receiving numerous awards, such as the W.R. Whitney Award of Advanced Materials Performance and Protection (AMPP) Society, the H.H. Uhlig Award of AMPP, the H.H. Uhlig Award of the Electrochemical Society (ECS), the U.R. Evans and T.P Hoar Awards from the British Corrosion Society, and is a technical fellow of four technical societies. He is an elected member of the Virginia Academy of Science, Engineering, and Medicine. He has developed and patented anti-microbial materials for high-touch surfaces that inactivate the COVID-19 virus and written several editorials in *CORROSION* discussing the "science behind it" associated with high-touch surfaces and copper impregnated masks for COVID-19 virus inactivation as well as lead in water and materials sustainability. He has investigated lead in drinking water piping systems to clarify materials issues central to disasters such as Flint, Michigan. He has served on numerous government review boards concerned with materials reliability, aging, and materials failures. These include the Columbia Accident Investigation Board, integrity of long-term storage canisters for spent nuclear fuel commissions in Canada, Sweden, Great Britain, as well as the United States. He also worked on the Defense Science Task Forces on Corrosion Control and Airborne Refueling. He has previously been a member of National Academies' corrosion studies such as ACE, ROCSE, and High-Performance Bolting Technology for Offshore Oil and Gas Operations.

SAWTEEN SEE provides consulting design services and is the president of See Robertson Structural Engineers. Ms. See was the managing director, Robert Bird Group (February 2019–January 2020); the design director, Robert Bird Group (June 2018–January 2019); and co-founded See Robertson Structural Engineers, LLC, with Leslie E. Robertson in 2018. She joined Skilling, Helle, Christiansen, Robertson in 1978 and co-founded Leslie E. Robertson Associates with Leslie Robertson in 1986, and was the managing partner from 1991 to 2017. She has extensive experience in the structural design of the full spectrum of building types. Particular expertise in tall building design, long-span structures; projects all over the world. She was the partner-in-charge of the structural engineering for the following: the 492m Shanghai World Financial Center; the 555m Lotte World Tower in Seoul; the 678.9m Merdeka 118 Tower, Kuala Lumpur, the second tallest building in the world; the Miho Museum and Bridge, Kyoto, Japan; Baltimore Convention Center Expansion; San Jose Convention Center; National Constitution Center, Philadelphia, Pennsylvania; NASCAR Hall of Fame and Museum, Charlotte, North Carolina; and the Rock & Roll Hall of Fame & Museum, Cleveland, Ohio; among others. She peer reviewed three buildings over 400m. She is a fellow, Institution of Structural Engineers, United Kingdom; member, National Academy of Construction; distinguished member and fellow, ASCE; fellow, New York Academy of Sciences; and received the following awards: Professional Achievement Award, Professional Women in Construction; and Asian Women in Business Leadership Award. She served on the advisory council for the School of Civil and Environmental Engineering at Cornell University and is currently an external examiner for the civil and structural engineering program at the University of Limerick, Ireland. She received a BSc (civil engineering) from Cornell University in 1977 and an MEng (structural engineering) from Cornell University in 1978.

HABIB TABATABAI is a professor of structural engineering and the director of the Structural Engineering Laboratory at the University of Wisconsin–Milwaukee. Prior to that, he was a principal structural engineer with the CTL Group (1989–1999) and a bridge engineer with the Florida Department of Transportation (1987–1989). He has conducted research on topics related to bridges, including reliability, durability, corrosion, and repair of structures, stay cable condition assessments, fatigue and vibration control, and survival analysis of structures. He wrote the National Cooperative Highway Research Program (NCHRP) Synthesis Report 353 on the inspection and maintenance of stay cables and served as the principal investigator (PI) on a Federal Highway Administration (FHWA) study on stay cables. He has also served as the co-PI for an NCHRP project on the condition evaluation of strands in concrete bridges and an FHWA study on jointless bridges. He is a licensed Structural Engineer and Professional Engineer in Illinois. He led the qualification testing of cables for all cable-stayed bridges built in the United States in the 1990s. He has published more than 120 technical publications and has a U.S. patent. He is a member of the Post-Tensioning Institute committee on cable-stayed bridges (DC-45) and has served on ASCE Committee 19 (Structural Applications of Steel Cables for Buildings) (2005–2020).

B

Information-Gathering Activities

JANUARY 24–25, 2022: OPEN SESSIONS (VIRTUAL)

Basis of Study National Science Foundation (NSF)
  Jim Ulvestad, Senior Advisor for Research Security, NSF
  Joy Pauschke, Ph.D., P.E., Program Director, Division of Civil, Mechanical and Manufacturing Innovation, Directorate for Engineering, NSF

Briefing from NSF on the NSF Study Grant Intentions
  Jim Ulvestad, Senior Advisor for Research Security, NSF

NSF Facility Oversight
  Matt Hawkins, Head, Large Facilities Office, NSF
  Linnea Avallone, Chief Officer for Research Facilities, NSF

Briefing from University of Central Florida (UCF)
  Francisco Cordova, Director, U.S. National Astronomy and Ionosphere Center, UCF

Arecibo Observatory Competition and Operations and Maintenance Transition
  Ralph Gaume, Division Director in the Division of Astronomical Sciences (AST), NSF
  Mike Wiltberger, Deputy Director, High Altitude Observatory, National Center for Atmospheric Research (NCAC), former Section Head of the Geospace Section in the Division of Atmospheric and Geospace Sciences, NSF

Failure Event Sequence
  Ashley VanderLey, Senior Advisor for Facilities, AST NSF
  Jim Ulvestad, Senior Advisor for Research Security, NSF

APPENDIX B

FEBRUARY 16–17, 2022: OPEN SESSIONS (VIRTUAL)

Briefing: NESC Arecibo Observatory Auxiliary M4N Socket Termination Failure Investigation
  Gregory Harrigan, Langley Research Center, Hampton, Virginia, United States
  Azita Valinia, Goddard Space Flight Center, Greenbelt, Maryland, United States
  Nathan Trepal, Kennedy Space Center, Merritt Island, Florida, United States
  Pavel Babuska, The Aerospace Corporation, El Segundo, California, United States
  Vinay Goyal, The Aerospace Corporation, El Segundo, California, United States

Briefing: Wiss, Janey, Elstner Associates, Inc., Arecibo Observatory Stabilization Efforts Analysis and Auxiliary Main Cable Socket Failure Investigation
  Jonathan C. McGormley, P.E., S.E., Principal, Wiss, Janey, Elstner Associates, Inc.

Perspectives on Grant Award and Operations of Arecibo Observatory Cooperative Agreement by UCF
  Ramon (Ray) Lugo III, former Director of UCF's Florida Space Institute and Principal Investigator of the NSF contract for Operation and Maintenance of the Arecibo Observatory
  Francisco Cordova, Director, U.S. National Astronomy and Ionosphere Center, UCF

Briefing: Arecibo Telescope Collapse—Forensic Investigation
  John Abruzzo, P.E., LEEP AP, Managing Principal and Forensics Practice Co-Leader, Thornton Tomasetti
  Liling Cao, Ph.D., P.E., LEED AP, Principal, Thornton Tomasetti

MAY 19–20, 2022: OPEN SESSIONS
SITE VISIT
ARECIBO OBSERVATORY, ARECIBO, PUERTO RICO

Observatory Tour and Inspection

Interview and Q&A with Engineering/Maintenance Staff

Supported Review of Existing Maintenance Records and Documentation On-Site, Particularly of the Cables and Sockets, and Drone Footage or Surveys of Other Cables

Post Site Inspection Q&A with Thornton Tomasetti in a Conference Room
  John Abruzzo, P.E., LEEP AP, Managing Principal and Forensics Practice Co-Leader, Thornton Tomasetti

OCTOBER 26, 2022: OPEN SESSIONS
BECKMAN CENTER, IRVINE, CA

Briefing: Thornton Tomasetti Arecibo Telescope Collapse Forensic Investigation
  John Abruzzo, P.E., LEEP AP, Managing Principal and Forensics Practice Co-Leader, Thornton Tomasetti

Perspectives and Discussions with Structural Engineers and Bridge Experts
  David Goodyear (NAE), Chief Bridge Engineer T.Y. Lin International (retired), Consulting Structural Engineer
  Karl H. Frank, Professor Emeritus, The University of Texas at Austin; Chief Engineer, Hirschfeld Industries (retired); and Consultant

C

Arecibo Telescope Cable Failure Mechanisms Considered by the Committee

Several material degradation phenomena produce a time-dependent loss in the cable load-carrying capability that can eventually trigger the failure of load-bearing cables that lack redundancy. Table 6.3-1 of the NASA Engineering and Safety Center report[1] summarizes these modes of time-dependent material degradation. The significance of these time-dependent degradation modes is that applied stresses well below the yield strength can activate or trigger certain phenomena leading to possible failure of materials in service even though the breaking strength or even a fraction of the yield strength was never exceeded in service by operative stresses from all sources. The relevance to the Arecibo Telescope is that factors of safety expressed by the minimum breaking strength of a cable in pristine condition divided by an applied stress, assuming the original load-bearing strength of the cable connection remains the same, do not apply to aged structures and cable connections with degraded load carrying capability. This is because the net section stress may rise as the connection material degrades by uniform thinning, creep, or crack propagation, which reduces the load-bearing capacity of the connection.

Moreover, the applied stress intensity factor at growing cracks may increase at the same or fixed global stress (or the same tension on wire cables) until the fracture toughness of the material is reached. This will depend on the size and growth of stress-concentrating flaws. Pull out of zinc discussed previously and referred to herein and by NASA, Thornton Tomasetti, Inc. (TT), etc., as "slip or flow" include time dependent creep and instantaneous plastic deformation. Based on the pristine original cable strength, the applied force or tension needed to cause creep is still only a fraction of the minimum breaking strength or yield strength achieved during one-time loading to fracture. In a creep failure process, fast loading eventually occurs as a sample cross-section of load bearing area decreases, and the remaining cross-sectional load-bearing area is the last material to separate by fast fracture. This fast fracture surface can exhibit a ductile mode of fracture, such as indicated by micro-scale micro void coalescence or macro-scale shear failure.

CREEP

The creep strain rate depends on the homologous temperature, T_H, given by absolute operating temperature relative to melting temperature ($T(K)/T_m(K)$), where K indicates Kelvin. Typical relationships have the following

---

[1] G.J. Harrigan, A. Valinia, N. Trepal, P. Babuska, and V. Goyal, 2021, *Arecibo Observatory Auxiliary M4N Socket Termination Failure Investigation*, NASA/TM–20210017934, NESC-RP-20-01585, NASA Engineering and Safety Center, NASA Langley Research Center, June, https://ntrs.nasa.gov/api/citations/20210017934/downloads/20210017934%20FINAL.pdf (hereafter "NESC Report").

APPENDIX C

form where n and m take on integer values or can be zero depending on the operative creep mechanism. The ratio of applied shear stress relative to the shear modulus is raised to the *n* power, giving a strong dependency on stress.

$$\frac{d\varepsilon}{dt} = A_i D_i \left[\frac{\sigma}{G}\right]^n \left[\frac{\sigma\Omega}{kT}\right] \left[\frac{b}{d}\right]^m$$

In this equation, A_i is a material constant, D_i is the diffusion rate for self-diffusion of zinc in this case, ε is strain, t is time, σ is shear stress, G is shear-modulus, k is Boltzmann's constant, T is temperature, b is the burgers vector of a dislocation, Ω is the atomic volume of the metal (zinc) undergoing creep, and d is the (zinc) grain size. In this expression, $n = 2$–6 in the case of power law creep (PLC) while m is often 0. There is, in this case, limited dependence on grain size. However, when Nabarro-Herring (N-H) or Coble creep prevails, $n = 0$ and $m = 2$–3. Hence, when creep occurs by these mechanisms, there is no dependence on stress as creep is driven by vacancy diffusion.

Thus, it is important to understand the prevailing creep conditions during the operation of the Arecibo Telescope. A material has different creep "regimes" based on the values of T/T_m and σ/G. as indicated on creep deformation maps.[2] Each material has its deformation map. In each regime, there is a dominant mode of creep such as Coble creep, Nabarro-Herring (N-H) creep, dislocation creep (or twinning), PLC, or dislocation glide instead of creep,[3] and the equation above is modified accordingly. These creep mechanisms are significant to the Arecibo Telescope failure as the creep regime dictates what factors matter and which do not. For instance, Coble and N-H creep do not depend on stress. N-H creep within sub-grains and power law creep depend not on grain size but are very sensitive to stress. NASA, TT, and the authors of this report agree that PLC was very likely operative in the zinc spelter socket at the Arecibo Telescope.[4,5] Given PLC, doubling the stress increases the creep strain rate substantially when $n = 4$ or 6 versus $n = 2$. More will be said about this below.

Despite all this information, the lifetime of components in service or laboratory coupons cannot be estimated from a deformation map. Textbooks note that low temperatures near 0.5 of T_m and moderate operating stresses, coupled with long service lifetime, represent the conditions where creep should be considered in engineering design. Unfortunately, in the Arecibo Telescope design, computing a static factor of safety based on static loads/stresses and materials properties with time-dependence degradation in their load-carrying capability was unfortunate. Load carry capability then would decrease over time in service.

It should be noted that deformation maps also report deformation regimes prevalent at low temperatures and high stresses where creep may not be operative and are short-circuited by dislocation glide. These convey the deformation regimes encountered near room temperature and at high load rates, such as during a tensile test. Consider a rising load tensile test over a short period. Here, deformation is elastic at low stress, and as stress is raised, dislocation glide occurs once the yield strength is exceeded. Time-independent plastic deformation occurs with possible work hardening until higher stresses are applied, which reach the theoretical breaking strength of the material. Creep is not operative in this regime. Fast laboratory loading at room temperature does not assess creep susceptibility or behavior during a creep process. Unfortunately, the laboratory tensile testing on replicated versions of failed parts falls in this category.

Furthermore, ramifications in a socket when creep of zinc enables load transfer amongst broomed steel wires is a complex consequence of creep not factored in when using routine safety factors. The spelter socket is normally not a weak point in the suspension system, but this is true only when under fast loading at room temperature, such as in a proof test. In the deformation map regime representative of a tensile test, elastic loading, the yield and breaking load dominated by time-independent elastic and plastic deformation of the material, is encountered. The following section turns attention back to the conditions and circumstances at Arecibo.

---

[2] H.J. Frost and M.F. Ashby, 1982, *Deformation-Mechanism Maps: The Plasticity and Creep of Metals and Ceramics*, Pergamon Press.
[3] T.H. Courtney, 1990, *Mechanical Behavior of Materials*, McGraw-Hill.
[4] TT Final Report.
[5] NESC Report.

The cast metal anchor concept for attachment of high-tensile wire cables relies on transferring wire tension forces to the zinc and the inner wall of the cast steel socket.[6] Annular, longitudinal, and radial compressive stresses are developed.[7] The adhesion and friction forces between the wire and cast zinc must be greater than the friction forces between the zinc and socket wall. In this arrangement, it is well-understood that initial zinc "flow" will occur during initial proof loading as zinc is engaged and seated. This is seen in proof testing, where wire failure must occur in the strand, not the socket. Bridge hanger design documents report that the socket should be designed to develop the ultimate strength without suffering measurable creep of the zinc under load. This stipulation seems to have been disregarded in the Arecibo Telescope design and the subsequent inspections by all the structural engineers.

The existing literature consistently observes that long-term cable socket slip under long-term loading is not just a matter of zinc engagement/repositioning within the socket as during initial loading. Detailed metallurgical training and analysis are not necessary to grasp these principles. For instance, Rehm, in the journal *Wire* in 1977,[8] wrote a subsection within the paper titled "Improving Slip and Creep Characteristics" where the second part of the subsection is labeled "the long-term slip which takes place after such repositioning if the stresses in the cone are still higher than the creep limit of the casting material." The paper goes on to explain possible methods for reducing slip. Vinet and Roberg[9] studied creep in creep-resistant Zn-Al alloys and showed the strong temperature, stress, and time inter-relationship enabling engineers to predict displacements on the order of 1 inch in 20 years at 30°C at 30 percent of the rated tensile strength for the poured metal cone-shaped socket.

In a component such as the spelter socket, permanent deformation of zinc changes the positions of wires during wire/core slippage and brings about zinc flow. Zinc flow can change the complex state of stress on the wires, such as the transfer of load to outer wires in the broom. The load transfer to the outer wires in the broom is exacerbated by non-uniformity in the broom. This is in complete agreement with NASA,[10] Wiss, Janney, Elstner Associates, Inc. (WJE),[11] and TT reports.[12] However, what is missing in these Arecibo Telescope socket failure mechanism investigations is a cogent explanation for the irregular, non-linear timeline of Arecibo Telescope's socket slip or core pullout. The mechanisms leading to Arecibo Telescope socket failure appear to have been arrested from 2004 to 2011, and perhaps even since initial seating in 1997. There is no evidence of increased socket pullout up to 2018, which appears to have then accelerated after Hurricane Maria, as well as after the M4N-T August 2020 cable failure and the M4-4 failure of November 2020.

The rapid increase in cable breaks is indirect evidence that supports the notion that slip was accelerated or the consequences were accelerated. The more noticeable consequences readily observed were the increase in wire breaks and the cracking in the back of socket M4-4T, as shown in Appendix E, Figure 29 of the TT Final Report.[13] It is possible that time-independent plastic deformation of an episodic nature occurred post-Maria, where instantaneous plastic flow was stimulated by an increase in tension. This plastic flow would have occurred abruptly on top of continual time-independent creep. The committee feels that both processes contributed to the overall failure and the accelerated pace of wire breaks just before the December 1, 2020, failure and the Arecibo Telescope collapse. The TT Final Report did include PLC of zinc, although its implementation in the finite element analysis of socket broomed wires is unclear.[14]

---

[6] I. Ridge and R. Hobbs, 2012, "The Behaviour of Cast Rope Sockets at Elevated Temperatures," *Journal of Structural Fire Engineering* 3(2):155–168. https://doi.org/10.1260/2040-2317.3.2.155.

[7] G. Rehm, M. Patzak, and U. Nurnberger, 1977, "Cast Metal Anchors for High-Tensile Wire Tendons," *Wire* 26(5):173–180.

[8] Rehm et al., 1977, "Cast Metal Anchors for High-Tensile Wire Tendons."

[9] R. Vinet and R. Roberge, 1978, "Evaluation of the Creep Resistance of Wire Rope End Attachments," *Journal of Engineering Materials and Technology* 100(2):214–216, https://doi.org/10.1115/1.3443475.

[10] NESC Report.

[11] Wiss, Janney, Elstner Associates, 2021, *Auxiliary Main Cable Socket Failure Investigation*, WJE No. 2020.5191, June 21 (hereafter "WJE Report"), p. 7.

[12] Thornton Tomassetti, Inc. (TT), 2022, *Arecibo Telescope Collapse: Forensic Investigation*, NN20209, prepared by J. Abruzzo, L. Cao, and P.E. Pierre Ghisbain, July 25, https://www.thorntontomasetti.com/sites/default/files/2022-08/TT-Arecibo-Forensic-Investigation-Report.pdf (hereafter "TT Final Report").

[13] TT Final Report.

[14] TT Final Report.

CORROSION

Corrosion of zinc, steel wire, and steel spelter socket casting is a thermodynamically spontaneous process that occurred naturally on Arecibo Telescope materials listed herein in the natural Arecibo environment. The issue is whether the corrosion rates had any effect on structural integrity. Material corrosion can occur in several modes, and atmospheric attack was likely the primary mode at Arecibo. Atmospheric corrosion of steel and zinc mainly occurs by a relatively uniform corrosion process, which sometimes may include shallow pitting. Corrosivity at Arecibo is controlled by environmental factors, the alloy chemistry, and any corrosion mitigation strategies applied. Corrosivity categories can be classified based on the environment.[15] Corrosion rates can then be estimated. Wet and dry cycling, time of wetness, relative humidity relative to salt deliquescence, salt deposition rates, dew point phenomena, and other atmospheric species, especially sulfur, affect corrosivity. Salt aerosols are often carried inland from seawater sources, and their presence depends on the sea state helping to create the aerosol, the distance of the structure in question from the ocean, wind conditions, time of wetness, as well as periodic rinsing by rainwater. NaCl and $MgCl_2$, as well as sulfates, are to be expected. SO_2 present in industrial applications is a potent factor.

Several publications define steel corrosion rates accurately as a function of atmospheric conditions using power law formulas expressing corrosion rate as a function of these factors.[16] The results for plain carbon steel would be modified by the wire steel composition.[17] Corrosion rates are modified by galvanizing.[18] Corrosion penetration results in a reduced wire cross-sectional area that progressively serves as the "remaining" load-bearing area after long exposure without mitigation. Hence, a fixed applied tension load and corrosive conditions would result in greater applied stress on the reduced cable net section over exposure time, gradually increasing toward a minimum breaking stress and, finally, failure. The safety factor is thereby eventually reduced to unity should this corrosion mode continue unabated. Failure by tensile overload could be achieved by corrosion at fixed tension should the load-bearing cross-section be reduced sufficiently such that minimum breaking strength is approached. However, this process is mitigated by galvanizing and application of a zinc-rich primer to the wires.

The scenario of reduced wire cross-section affecting structural integrity was found to be unlikely at Arecibo. The spelter socket casting was too thick for any appreciable change in the load-bearing cross-section of the socket itself. Moreover, wires were well protected by zinc galvanizing, as well as a zinc-rich primer, and the sockets had humidity controls. The zinc-rich primer was periodically reapplied. The maintenance activities at the Arecibo Telescope were effective in this regard. The zinc functions as a barrier and a sacrificial anode to protect any steel wires exposed.[19] Little zinc corrosion was observed in failed spelter socket cross-sections. Hence, any scenario placing uniform corrosion as the source of failure is extremely unlikely at Arecibo due to insufficient corrosion penetration and minimal loss of load-bearing areas on each subcomponent.[20]

The zinc-rich primer and galvanizing (ASTM Standard 586; class A, 1 oz/ft^2 depending on wire diameter[21] with zinc subject to specification B 6) was fairly intact at the Arecibo Telescope, meaning that this sacrificial cathodic protection was still operative at the time of failure and likely protected the steel from the aforementioned hypothetical loss in cross-section. One ounce/ft^2 can protect to the point that no rusting is observed for more than 10 years in a marine environment and 20 years in a rural environment.[22] The exact protection in the tropical environment of Arecibo is uncertain. Failed wires were reported to have from 0.85–1.35 ounces/ft^2 of galvanized zinc.[23] White corrosion products traced to zinc carbonates were observed, indicating the zinc was corroding, while

---

[15] International Organization for Standards (ISO), 2012, "ISO 9223: Corrosion of Metals and Alloys—Corrosivity of Atmospheres—Classification, Determination and Estimation," https://www.iso.org/obp/ui/#iso:std:iso:9223:ed-2:v1:en.

[16] W. Hou and C. Lang, 2004, "Atmospheric Corrosion Prediction of Steels," *CORROSION* 60(3):313–322.

[17] ANSI, 2020, "ASTM G101: Standard Guide for Estimating the Atmospheric Corrosion Resistance of Low-Alloy Steels."

[18] J.W. Spence. F.H. Haynie, F.W. Lipfert, S.D. Cramer, and L.G. McDonald, 1992, "Atmospheric Corrosion Model for Galvanized Steel Structures," *CORROSION* 48(12):1009–1019, https://doi.org/10.5006/1.3315903.

[19] D.A. Jones, 1995, *Principles and Prevention of Corrosion*, 2nd Edition, Pearson.

[20] TT Final Report.

[21] American Society for Testing and Materials (ASTM), 1998, "ASTM A586-98: Standard Specification for Zinc-Coated Parallel and Helical Steel Wire Structural Strand and Zinc-Coated Wire for Spun-In-Place Structural Strand," https://doi.org/10.1520/A0586-98.

[22] A.J. Stavros, 1987, "Corrosion," *AMS Metals Handbook*, Ninth Edition, Volume 13, ASM International, p. 432.

[23] TT Final Report.

very little red rusting was seen, which likely indicates steel corrosion.[24] Zinc corrosion products are indicative of a functional cathodic protection strategy. The observed presence of actively corroding zinc suggests ongoing functional cathodic protection of the steel. Under cathodic polarization, the steel corrosion rate could be reduced by a factor of up to 100. The presence of zinc galvanizing and zinc primer mitigated corrosion of the steel to the extent that reduction in a load-bearing cross-section was unlikely. However, zinc presents a "trade-off" and leads to enhanced hydrogen production from the reduction of water, which can affect hydrogen-assisted cracking susceptibility. This is discussed below.

STRESS CORROSION CRACKING AND HYDROGEN-ASSISTED CRACKING

Steel wire utilized as tension elements in cables of suspension bridges and at the Arecibo Telescope brings together all the factors necessary to render them susceptible to hydrogen embrittlement (HE), more specifically, hydrogen environment-assisted cracking (HEAC) and stress corrosion cracking. Stress corrosion cracking (SCC) is the cracking of a material produced by the combined action of corrosion and sustained tensile stress.[25] To further illustrate the insidious nature of SCC, it should be noted that cracking can initiate and grow at stresses well below the yield strength, rendering a safety factor based on stress not useful. If a crack-like defect exists in a steel cable, which places the applied stress intensity factor (SIF) below the dry air fracture toughness but above the threshold SIF for SCC initiation and crack growth, failure may occur by fast fracture after a period of slow SCC growth, which reduces the load-bearing cross-sectional area. SCC susceptibility requires a susceptible material governed by composition/microstructure, applied tensile stress (flaw-free) or SIF (flaw present), and a corrosive environment. SCC and HEAC are thus material-specific. Steel susceptibility is a function of alloy composition, microstructure, and mechanical properties.

High-strength steels with high material hardness are more susceptible than lower-strength alloys.[26] The cable wires in this application may be regarded as high strength, as indicated by their hardness. The stress may come from multiple sources and may be residual self-equilibrated stresses contained in the part or gravity-load induced. Removal of the tensile stress, the detrimental metallurgical condition, or the corrosive environment eliminates the conditions where SCC is operative. None of these is possible at the Arecibo Telescope.

SCC also describes hydrogen-assisted cracking, where hydrogen is produced due to corrosion. HE, when exposed to a corrosive environment, is also called SCC. In the NASA report, HE is termed HEAC or hydrogen-assisted cracking. The ASTM definition of HE states that it is cracking in the presence of hydrogen. In this application, hydrogen is produced electrochemically from water and proton reduction during corrosion of the steel wire and/or by cathodic protection when steel is protected by zinc. Steel wires at 200 ksi strength levels, such as cold-drawn UNS G10800 steel, are susceptible to HE at electrode potentials experienced when galvanically coupled to zinc, which is documented by Enos and the references within.[27] A portion of the hydrogen generated is absorbed into the steel, which lowers the breaking load, tensile strain, and threshold stress intensity in case of pre-existing flaws that concentrate stress. Under static tensile loads, cracks can initiate and grow at stresses well below the yield strength. The breaking strength and ductility of steels are reduced by the hydrogen concentration developed in the alloy. HEAC depends on the hydrogen concentration absorbed and is often marked by a critical hydrogen concentration where the fracture mode changes from ductile to brittle. The breaking strength and ductility of steels are reduced by the local hydrogen concentration developed in the alloy. HE, and the broader category of SCC, is plastic strain-rate dependent. Fast strain rates would deny the chance for brittle fracture even if hydrogen is pre-charged, while slow strain rates or static loading in high-strength alloys allow it to operate. This is because hydrogen must be repartitioned by solid-state diffusion (which is slow) to the high tri-axial stress at the crack tip. This goes hand-in-hand with observing ductile fracture in wires experiencing overload and fast fracture during rapid loading.

---

[24] TT Final Report.

[25] ASTM, 1993, *Annual Book of ASTM Standards: Wear and Erosion; Metal Corrosion*, Vol. 03.02, ASTM International.

[26] Stavros, 1987, "Corrosion," p. 432.

[27] D.G. Enos and J.R. Scully, 2002, "A Critical-Strain Criterion for Hydrogen Embrittlement of Cold-Drawn, Ultrafine Pearlitic Steel," *Metallurgical and Materials Transactions A* 33:1151–1166. https://doi.org/10.1007/s11661-002-0217-z.

A fraction of the wires in the spelter socket outside the core were noted to fracture in a brittle fashion by the slow process described above. However, fast fracture at a high strain rate in the laboratory during a fast tensile test is not a viable methodology to detect SCC or HE susceptibility. The particular morphology of fracture surfaces observed at high magnification are often fingerprints or indicators of cracking mode. Fatigue, SCC, and hydrogen-assisted cracking have their own telltale flat fracture morphologies, which may be microscopically seen as cleavage or intergranular. Brittle failure of prestressing wires can also occur by longitudinal splitting.[28]

Arecibo Telescope steel wire of 220 ksi tensile strength subject to ASTM A 586 was utilized.[29] The steel met the strength ductility of drawn AISI 1080 or UNSG10800 steel. This steel is near the eutectoid composition. Hence, the microstructures developed ultra-fine eutectoid pearlite. One hundred and sixty-eight wires were assembled using a helical arrangement. At the Arecibo Telescope, steel wires underwent fracture in the socket—but outside the wire broom in one failure and outside the zinc casting but near the socket in another case—during the final collapse.

Broken wires have been known and reported occasionally.[30] Subcritical breaks occurred in many of the Tower 4 wires outside the socket core but at random positions with respect to the socket core. Whether inside or outside, steel wires were exposed to zinc and periodic moisture. Concerning HE, zinc facilitates water reduction and hydrogen entry, and the hydrogen centration in the steel is greater than what can be introduced by atmospheric corrosion alone without zinc. This is due to cathodic polarization brought about by the zinc galvanizing and zinc-rich primer in a galvanic couple with the steel. The zinc potentially increased the hydrogen content of the wires at the Arecibo Telescope. Thus, the HEAC should have been possible, but little evidence was observed. Several wires underwent slow crack growth before the final fast fracture of the remaining wires. This is indicated from cross-section examinations and was indicated by black/brown corrosion products, which suggest the cracks found predate the actual failure event. Moreover, subcritical cracks away from the primary fracture were detected upon inspection of a few wires that failed during the collapse of Tower 4.

HE should likely be removed from consideration as a primary root cause of the failure. This is because brittle cracking was indicated by fractography on only a small fraction of the wires in the outer region of the broomed wires and anywhere else. Subsequently, Lehigh University confirmed in Fritz laboratory tests that the breaking strength of steel cables and sockets during fast fracture was not degraded during laboratory tensile tests of failed sockets/cables. The Phoenix et al. paper indicates the same.[31] However, it should be noted that this test was conducted at moderate to fast loading rates, which would not have detected the presence of HEAC due to its slow strain rate dependence. When it is stated that the steel wires were not weakened during service, that is a correct statement in the case of a tensile test applied during fast loading at room temperature, such as in a proof test. Fractography inspections indicate mainly ductile cup/cone failure and shear wire breaks versus HEAC. This is the expected result in the case of a final fast fracture in an overload situation at the time of the collapse. This supports the ductile wire separation argument and points elsewhere to find a primary root cause. Other arguments, analyses, and evidence are all in favor of zinc slip pull-out and zinc rupture. Hydrogen content was not measured in the steel, nor were slow strain tests conducted on harvested wires when galvanically coupled to zinc. Therefore, the overall extent of HE damage to wires is currently unknown, but the lack of fractographic evidence does not support HE as the root cause of failure.

FATIGUE

Steel wires and cables are susceptible to fatigue. Fatigue leads to failure at applied static tensile stress well below the yield point in uniaxial tension or under mixed loading modes. A small cyclic tensile stress amplitude well below the yield point but sufficiently high (i.e., above the endurance limit) triggers crack initiation and propagation in a ferrous material given enough cycles. Fatigue has three stages. These are initiation, propagation, and growth until fast fracture occurs and is brought about by a reduced remaining load-bearing cross-section due

---

[28] Ibid.
[29] TT Final Report.
[30] L. Phoenix, H.H. Johnson, and W. McGuire, 1986, "Condition of Steel Cable After Period of Service," *Journal of Structural Engineering* 112(6):1264, https://doi.org/10.1061/(ASCE)0733-9445(1986)112:6(1263).
[31] Phoenix et al., 1986, "Condition of Steel Cable After Period of Service."

to fatigue crack growth. Fatigue depends on the material, and a specific material's behavior is often depicted in a plot showing maximum applied cyclic stress causing initiation, propagation, and growth as a function of the number of cycles of the load. Greater cyclic load amplitudes reduce the number of cycles to failure. In steels, the endurance limit is defined as the applied cyclic stress below which there are an infinite number of cycles needed to attain initiation and propagation. Given enough cycles, fatigue failure will occur at a lower global stress below the minimum break strength observed in a single loading to failure. Fatigue can be affected by corrosion either by changing initiation, propagation, and/or growth. Corrosion followed by fatigue test in dry air allows sampling of the effect of corrosion on fatigue, which produces surface damage and increases stress concentration. True "corrosion fatigue" is concurrent corrosion and fatigue, which is usually transgranular and indicated by markings (striations) related to crack advance as a function of number of cycles.

The potential sources of fatigue loading are numerous at the Arecibo Telescope and include vibrations, wind loading, and even day/night cycling, given applied tie-down tension and thermal expansion and contractions. Oscillations were partially mitigated with the use of dampening devices. Fatigue loading cycles tend to have a small cyclic stress amplitude relative to static stress. This level of cyclic stress has been regarded to be insufficient to trigger fatigue damage in this application. This was also concluded previously.[32] The location of the fracture inside the spelter socket also argues strongly against a fatigue failure mode. At the Arecibo Telescope, fracture surfaces did not indicate fatigue as far as the usual accompanying observation of striations or markings on fracture surfaces. Fatigue striations were not observed at the Arecibo Telescope.

LONG-TERM ELECTROPLASTICITY

Troitskii and Likhtman[33] introduced modern electroplasticity. Zuev et al.[34,35] discussed the mobility of dislocations in zinc single crystals under the action of electric current pulses. Stashenko et al.[36] reported the significant effect of current pulses on the creep of zinc single crystals during short-term laboratory tests involving high current densities. More recent papers by Conrad,[37] Guan et al.,[38] Lahiri et al.,[39] and Rudolf et al.[40] address the use of electroplasticity in manufacturing (also known as electrically assisted manufacturing).

Stashenko et al.[41] reported that the "flow of conductivity electrons gives part of its energy to defects which participate in plastic deformation of the metal." The authors concluded that the "pulse action of the current on zinc single crystals during creep is accompanied by a considerable increase in the creep rate and by a discontinuous increase in deformation." Conrad reported that electric and magnetic fields can often have a significant effect on the plastic deformation of metals and ceramics.[42] Conrad performed experiments on the effects of external DC

---

[32] Ibid.

[33] O. Troitskii and V. Likhtman, 1963, "The Effect of the Anisotropy of Electron and γ Radiation on the Deformation of Zinc Single Crystals in the Brittle State," *Doklady Akademii Nauk SSSR* 148:332–334.

[34] L.B. Zuev, V.E. Gromov, and V.F. Kurilov, 1978, "Mobility of Dislocations in Zinc Single Crystals Under the Action of Electric Current Pulses," *Doklady Akademii Nauk SSSR* 239(1):84–86 (in Russian).

[35] L.B. Zuev, V.E. Gromov, and L.I. Gurevich, 1990, "The Effect of Electric Current Pulses on the Dislocation Mobility in Zinc Single Crystals," *Physica Status Solidi (a)* 121(437).

[36] V.I. Stashenko, O.A. Troitskii, and V.I. Spitsyn, 1983, "Action of Current Pulses on Zinc Single Crystals During Creep," *Physica Status Solidi (a)* 79(549).

[37] H. Conrad, 1998, "Some Effects of an Electric Field on the Plastic Deformation of Metals and Ceramics," *Materials Research Innovations* 2(1):1–8, https://doi.org/10.1007/s100190050053.

[38] L. Guan, G. Tang, and P.K. Chu, 2010, "Recent Advances and Challenges in Electroplastic Manufacturing Processing of Metals," *Journal of Materials Research* 25(7):1215–1224.

[39] A. Lahiri, P. Shantraj, and F. Roters, 2019, "Understanding the Mechanisms of Electroplasticity from a Crystal Plasticity Perspective," *Modelling and Simulation in Materials Science and Engineering* 27:085006.

[40] C. Rudolf, R. Goswami, W. Kang, and J. Thomas, 2021, "Effects of Electric Current on the Plastic Deformation Behavior of Pure Copper, Iron, and Titanium," *Acta Materialia* 209:116776.

[41] Stashenko et al., 1983, "Action of Current Pulses on Zinc Single Crystals During Creep."

[42] Conrad, 1998, "Some Effects of an Electric Field on the Plastic Deformation of Metals and Ceramics."

electric fields during the superplastic deformation of a 7475 aluminum alloy. According to the author, the surface charge reduced the flow stress by 10–20 percent.[43]

Kim et al.[44] discussed the origins of electroplasticity in metallic materials. They reported that, in a system that includes a grain boundary (i.e., general defect in polycrystalline metallic materials), charge imbalances near defects would "drastically weaken atomic bonding under electric current." The authors also note, based on tests on magnesium and aluminum alloys, that "the weakening of atomic bonding was confirmed by measuring the elastic modulus under electric current, which is inherently related to the atomic bonding strength." Xu et al.[45] performed tensile tests on dog-bone specimens made with a Mg-3Al-1Sn-1Zn alloy. Current pulses with peak current densities of 0, 20, and 30 A/mm^2 were applied to the specimens during tests. Based on stress-strain diagrams, the authors reported a drop in flow stress and increased fracture strain as the current densities increased. They reported that significant dynamic recrystallization developed with a current density of 30 A/mm^2.

The flow of electrons in the cables caused increased temperature due to Joule heating. Stashenko et al.[46] note that the "skin effect pressing a high-frequency current back to the outer areas of the sample … may result in an overheating of the surface." Although creep is highly dependent on temperature, the Joule effect and the temperature rise due to the skin effect may not be significant compared to the potential electroplastic effects.

---

[43] Ibid.

[44] M.J. Kim, S. Yoon, S. Park, H.-J. Jeong, et al., 2020, "Elucidating the Origin of Electroplasticity in Metallic Materials," *Applied Materials Today* 21, https://doi.org/10.1016/j.apmt.2020.100874.

[45] H. Xu, Y.-J. Zou, Y. Huang, P.-K. Ma, Z.-P. Guo, Y. Zhou, and Y.-P. Wang, 2021, "Enhanced Electroplasticity Through Room-Temperature Dynamic Recrystallization in a Mg-3Al-1Sn-1Zn Alloy," *Materials* 14(3739), https://doi.org/10.3390/ma14133739.

[46] Stashenko et al., 1983, "Action of Current Pulses on Zinc Single Crystals During Creep."

D

Arecibo Telescope Design Issues Considered by the Committee

The committee considered the following issues for design wind speed:

- Wind speeds vary with time and elevation but generally increase with height above ground. Design wind speeds used before the publication of ASCE 7-95 (1995) were measured differently compared to the wind speeds used in current standards. The older standards measured wind in terms of "fastest mile." With ASCE 7-95 and later editions, the basis of design wind speed was changed to "3-second gust" speed; both measured at 33 ft above ground for the Exposure C category.
- Example wind design speeds specified by design standards:
  - ASCE 7-22, Table C26.5-7: 76 mph "fastest-mile"
  - ASCE 7-93 and prior: 90 mph "3-second gust"
  - ASCE 7-95 through 7-05: 115 mph "3-second gust"
- Per ASCE 7-22 (ASCE 7-22 Figure 26.5-1C and Figure 26.5-1D), design wind speeds for Puerto Rico (and Hawaii and U.S. Virgin Islands) shall be determined from the ASCE Wind Design Geodatabase, which can be accessed at the ASCE 7 Hazard Tool[1] or approved equivalent. The relevant design wind speeds for Building Risk Category III and IV are shown in ASCE 7 Hazards Reports.[2] Per ASCE 7-22, the ultimate wind speeds for the Arecibo Observatory's location are 318 mph and 334 mph for Risk Category III and IV, respectively; these are to be used with a load factor of 1.0.
- In ASCE 7-22 (Minimum Design Loads and Associated Criteria for Buildings and Other Structures (ASCE/SEI 7-22), the design wind speeds are determined based on a building's Risk Category (I through IV). The building or structure's Risk Category depends on the nature of its occupancy (see ASCE 7-22 Table 1.5-1). Per ASCE 7-22, the Arecibo Observatory should be classified as Category IV: "Buildings and Other Structures Designated as Essential Facilities."
- The wind speed's mean recurrence interval (MRI) implied for Risk Categories I through IV are 300, 700, 1,700, and 3,000 years, respectively. The wind provisions of ASCE 7-22 specify 1,700-year ultimate wind loads directly for Risk Category III buildings (ASCE 7-22 Figure 26.5-1C) and 3,000-year ultimate wind loads for Risk Category IV buildings (ASCE 7-22 Figure 26.5-1D), both with load factor = 1.0.

---

[1] The ASCE 7 Hazard Tool is available at https://asce7hazardtool.online, accessed April 11, 2023.
[2] See the following ASCE 2023 Standards: "Standards ASCE/SEI 7-22—Risk Category III" and "Standard ASCE/SEI 7-22—Risk Category IV," both available at https://asce7hazardtool.online, accessed April 11, 2023.

APPENDIX D

- The design wind speeds are associated with load and resistance factor design (LRFD) and allowable stress design, which have changed with changes in basic wind speeds. Currently, 1,700-year MRI ultimate wind loads are used with a load factor of 1.0 in the LRFD load combinations. In the past, 100-year MRI strength design-based wind speed was used along with a load factor of 1.6. The results from these two approaches are somewhat similar and consistent with best international practice.
- In the Thornton Tomassetti (TT) report *Arecibo Telescope Collapse: Forensic Investigation*, Appendix J, page 6, TT wrote on Hurricane Maria's wind speeds: "We also note that the maximum instantaneous wind speed recorded during the storm is 108 mph, or two percent less than the second upgrade's design wind speed of 110 mph."[3] However, TT's Appendix J Figure 6 showed that the peak speed of 108 mph was measured over 10 minutes (the red plot), and the instantaneous wind speed (the black plot) was measured every 15 seconds.
- Hurricane Maria occurred in September 2017. In 2017, design wind speeds were based on 3-second gust wind and not on maximum instantaneous wind speed. The second upgrade was in 1992 and may/could have used "fastest mile wind" then.
- Therefore, comparing the maximum instantaneous wind speed of 108 mph in 2017 to the second upgrade's design wind speed of 110 mph (1992) is not technically correct.

It would have been more meaningful for the engineers/consultants to use the original design wind speed, codes, etc., and compare it to designs using the design wind speed and codes for the first and second upgrades, and lastly, to designs using the current design wind speed and codes. That would have been a useful and proper comparison to make. All the various write-ups of wind speeds without their qualifiers and basis of measurements in the various engineers/consultants' reports are not technically meaningful nor correct. Codes and standards are excellent tools, but to select a criterion, such as the wind speed, from one code and apply it to another code, decades more current, is wrong. These documents often have calculation procedures, variables, and look-up tables that are intended to provide a reliable methodology but are not transparent with respect to the applicable physics and engineering theory.

Thornton Tomassetti, Inc. (TT), 2022, *Arecibo Telescope Collapse: Forensic Investigation*, NN20209, prepared by J. Abruzzo, L. Cao, and P.E. Pierre Ghisbain, July 25, https://www.thorntontomasetti.com/sites/default/files/2022-08/TT-Arecibo-Forensic-Investigation-Report.pdf.

E

Acronyms and Abbreviations

A&W	Ammann & Whitney (engineering consultants)
AASHTO	American Association of State Highway and Transportation Officials
AO	Arecibo Observatory
ARPA	Advanced Research Projects Agency
ASCE	American Society of Civil Engineers
ATM	Division of Atmospheric Sciences (NSF)
CW	continuous wave
EIRP	Effective Isotropic Radiated Power
EM	electromagnetic
EOR	Engineer of Record
EP	electroplasticity
FE	finite element
FMEA	failure modes and effect analysis
FMECA	failure modes effects and criticality analysis
FY	fiscal year
HAC	hydrogen-assisted cracking
HE	hydrogen embrittlement
HEAC	hydrogen environmentally assisted cracking
LEP	low-current electroplasticity
M4-2T	Main Cable 2 Tower 4 (third cable failure)
M4-4T	Main Cable 4 Tower 4 (second cable failure)
M4N-T	Auxiliary Cable North Side Tower 4 (first cable failure)
MPE	maximum permissible exposure

O&M	operations and maintenance
OMB	Office of Management and Budget
NAIC	National Astronomy and Ionosphere Center
NASA	National Aeronautics and Space Administration
NESC	NASA Engineering and Safety Center
NSF	National Science Foundation
PLC	power law creep
PTI	Post-Tensioning Institute
RF	radio frequency
RFP	request for proposal
SCC	stress corrosion cracking
SRI	SRI International
TT	Thornton Tomasetti, Inc. (engineering consultants)
UCF	University of Central Florida
WJE	Wiss, Janney, Elstner Associates, Inc. (engineering consultants)
WSP	WSP Global, Inc.